Nural Öztürk
Sibel Karaca
Editors

Triage in Radiation Accidents

nova
Medicine & Health
New York

DOI: https://doi.org/10.52305/GOBE1802

NOTICE TO THE READER

Library of Congress Cataloging-in-Publication Data

ISBN: 979-8-89530-415-0 (Hardcover)
ISBN: 979-8-89530-451-8 (eBook)

Published by Nova Science Publishers, Inc. † New York

Contents

Preface

Radiation has existed naturally since the formation of the world, but it has also been obtained artificially as a result of curiosity and research. After radiation was obtained artificially, it became a part of daily life and began to be widely used in many disciplines such as health, industry and energy production. Radiation has extensive applications worldwide, especially in the field of health. While radiation was initially thought to have only beneficial applications, it has become clear over time that it can be dangerous. With the widespread use of radiation, catastrophic radiation accidents have begun to occur at an increasing rate due to technical errors or user errors. It has been observed that radiation can pose non-negligible risks for both the public and employees in the event of a possible accident.

Assistance to radiation victims after radiation accidents caused by various reasons is an important issue. The way a radiation managing accident occurs is of vital importance for both the victims and those who intervene. The key to successfully managing a crisis that may occur during a radiation accident is to examine the possible dangers before the accident and take the necessary precautions. Creating crisis scenarios and determining the personnel who will take part in these scenarios in advance are included in solution-oriented approaches. Task definitions, equipment preparation and personnel training within the scope of the scenario will ensure that the possible disaster is overcome with the least damage. Radioactive contamination must be brought under control in possible radiation accidents. It is not an easy process to fully control radioactive contamination. If a panic environment is created due to a lack of planning and organized action is not taken, the victims may receive serious damage or die. In addition, those who intervene may be exposed to high doses of radiation. Therefore, if the necessary precautions are not taken before a possible radiation accident, the accidents may reach catastrophic dimensions. The most important steps to be taken to prevent pre-determined dangers are the necessary plans and informed and trained personnel must be ready to take action in a timely manner.

The idea of writing this book emerged as a result of the determination that there was not enough literature on the procedures to be applied in the event of a radiation accident or a radiation leak. In addition, there is no sourcebook that provides comprehensive information on the measurement and grading of radiation leakage, the definition of its physical and psychological effects, and the precautions to be taken to prevent possible accidents.

The content of this book includes basic concepts related to radiation. Different scenarios and applications for radiation protection are also included in the scope of the book. Understanding the radiobiological effects of radiation on living tissue is one of the main topics covered in the book. In this book, triage priorities and rules in the event of any radiological or nuclear accident are explained in detail. In addition, this book examines major radiation accidents that have occurred in different parts of the world. It also focuses on radiation-related infections that may occur after radiation accidents and the neurological and psychological effects of radiation accidents on children and adults.

This book is a practical educational resource with useful information for medical professionals (especially emergency physicians, oncologists, haematologists, neurologists, etc.), medical and physics students, nuclear and health physicists, academics, researchers and all other relevant professionals. The most important contribution of this book is the recommendation for what should happen in the event of a possible radiation accident and the rules of intervention.

About the Editors

Nural Öztürk, PhD
Assistant Professor

Nural Öztürk earned her bachelor's degree from the Physics Department, Faculty of Science at Istanbul University, and earned a master's degree from the Department of Medical Radiophysics of the Health Institute at Istanbul University. Between 1991-1995, she worked as a specialist at the Department of Oncology, Dokuz Eylül University in Izmir, Turkey, and between 1995-2017, she worked at the Radiation Oncology Clinics of the Ministry of Health in Istanbul. She received her PhD degree from the Department of Medical Radio Physics of the Health Institute at Istanbul University in 1999. In 2017, she became a lecturer for the Department of Radiation Oncology at Aydın Adnan Menderes University, Turkey. Her areas of interest in clinical applications are pediatric axial irradiation techniques, patient-specific breast Ca techniques, 3D gynaecological brachytherapy applications, patient/device quality control and device beam modelling applications. She has co-authored various publications, covering topics of medical physics applications, brachytherapy quality control and radiation oncology. She has also written chapters on medical physics for local and international scientific media.

Sibel Karaca, PhD
Associate Professor

Sibel Karaca earned her master's degree from the Radiation Oncology Depertment at Atatürk University, Health Sciences Institute. She worked as a radiotherapy physicist at the Radiation Oncology Clinics of Ministry of Health Hospitals between 2006 and 2020. After completing her PhD, she became an Associate Professor of Health Physics in 2023. Dr. Karaca is currently working as an Associate Professor in the Department of Radiation Oncology at Akdeniz University, Antalya, Turkey. She has conducted dosimetric and clinical studies, especially in the field of radiotherapy physics. Her other work focuses on radiation education. She has published articles in well-known medical, physics and radiation oncology journals as first author and co-author.

Chapter 1

Radiation Basics

Nurdan Özbek[1,2,*] and Nural Öztürk[3]

[1]Ümraniye Training and Research Hospital Radiation Oncology Clinic, Department of Physics, Ümraniye, İstanbul, Türkiye
[2]İstanbul University-Cerrahpaşa, Cerrahpaşa Faculty of Medicine, Graduate Education Institute, İstanbul, Türkiye
[3]Department of Radiation Oncology, Aydın Adnan Menderes University, Aydın, Türkiye

Abstract

Atom is derived from the Greek word "atomos," which means indivisible. The first person to use the word atomus was Democritus, who lived in the 440s BC. An atom is the smallest building block that carries the chemical properties of an element. It basically consists of an atomic nucleus and a cloud of electrons around it. Radioactivity is the decay of the atoms of some substances and the emission of particles or energy. After the discovery of radioactivity by the French physicist H. Becquerel in 1896, Marie Skłodowska Curie discovered the radioactive property of thorium as a result of her experiments with uranium and isolated the element radium. As a result of these studies of radioactivity, rapid progress was made in the concept of radiation. Radioactive elements are divided into two groups: natural and artificial. These elements emit excess energy as ionised radiation to achieve a temporary or permanent equilibrium. Ionised radiation is a type of radiation that has enough energy to knock electrons out of the orbits of atoms. Charged particles cause direct ionisation, while photons and neutrons cause indirect ionisation. Radiation is used for many purposes, particularly in health, industry, research and nuclear reactors. The

* Corresponding Author's Email: nurdan.ozbek@gmail.com.

In: Triage in Radiation Accidents
Editors: Nural Öztürk and Sibel Karaca
ISBN: 979-8-89530-415-0

purpose of this section is to provide basic information about radiation, starting with the definition of radiation.

Keywords: atom, decay, ionisation, radiation, radioactivity

1. Introduction

The word atom, first used by Democritus, who lived in the 440s BC, is derived from the word "atomic," which means indivisible. The atom, the smallest building block of objects in the universe, consists of protons, neutrons and negatively charged electrons orbiting around a positively charged nucleus. Protons, which make up the basic structure of the atom, are positively (+) charged, electrons are negatively (-) charged and neutrons are uncharged (Khan et al., 2003). Therefore, the electrical charge of the atom is determined by the number of protons and electrons, and the number of protons determines the nuclear charge.

2. Atom and Molecular Properties

The mass of the nucleus, which occupies a small space in the centre of the atom, is greater than the mass of the electrons in the surrounding layers. Since the mass of the electron is negligibly smaller than the mass of the nucleus, the mass of the atom is approximately equal to the mass of the nucleus. Particles that interact with each other through the electric force of attraction between charged particles, protons and electrons, come together to form an atom. There is a nuclear force of attraction between protons and neutrons.

Looking at the fundamental particles of the atom, the electrons are represented by the symbol "e" and their mass is 9.1094x10 -31 kg (kilogram). They revolve around the nucleus, both around themselves and in their orbits according to their energy level. The orbits around the nucleus carry a certain number of electrons, and each orbit is called an electron orbit. The innermost orbital (shell) is called K, the others are called L, M, N, and O… respectively. The maximum number of electrons that can be found in the shell, $2n^2$, is determined by using the principal quantum number (n), which has values of n=1,2,3,4.5,… respectively (Khan et al., 2003). While the principal quantum number indicates the distance from the nucleus, as the number gets smaller, the distance to the nucleus decreases, and the binding energy increases.

The electrons in the outermost orbital are valence electrons and the number of valence electrons determines the chemical properties of the atom. Each orbital represents a different energy level. The minimum energy level in the first orbit is called the ground state, and in this state, when electrons gain energy from external influences, they rise to higher energy levels. Electron energy can be found by adding kinetic energy and electrical potential energy, and this energy increases as the number of orbital increases. This shows that the energy level of the electron increases as it moves away from the nucleus. The amount of energy required to release the electron from its orbital is called the binding energy. The ionisation energy is the binding energy of the electron in the outermost orbit of the atom. The symbol for the proton is "p" or "H +" and its mass is 1.6726 x 10 -27 kg (kilogram), so it has about 1836 times the mass of the electron. The number of protons is an important factor in determining the physical and chemical properties of the atom (Khan et al., 2003).

- The mass of the neutron is approximately equal to the mass of the proton. The symbol for the neutron is "n" and it has no charge (Khan et al., 2014). Since atoms are the smallest building blocks, the atomic mass unit (amu), a unit of measurement used to calculate the mass of molecules, is used in mass measurement. It is expressed as 1 amu = 1.66x10 -27 kg (kilogram). Using this equation, the electron is 0.000548 amu, the proton is 1.00727 amu and the neutron is 1.00866 amu (Kamal, 2014). Under normal conditions, atoms have an equal number of positively charged protons and negatively charged electrons and are neutral. Depending on whether the atom gains or losses electrons after ionisation, the proton-electron balance are disturbed, and the atom becomes negatively or positively charged and becomes an ion.
- While the difference between the number of protons and electrons indicates the charge of the atom, the mass number indicates the sum of the number of protons and neutrons, and the atomic number indicates the number of protons. Atoms are classified according to the number of neutrons and protons (Maher, 2011).
- While the difference between the number of protons and electrons shows the charge of the atom, the mass number shows the sum of the number of protons and neutrons, and the atomic number shows

the number of protons. Atoms are classified according to the number of neutrons and protons (Maher, 2011).

- Isotope atoms have the same number of protons and electrons but different numbers of neutrons, which gives them the same physical and chemical properties.
- Atoms with the same total number of protons and neutrons, that is, the same mass number, but different numbers of neutrons and protons, are called isobar atoms. These atoms have different chemical and physical properties.
- Isotone atoms are atoms with the same number of neutrons but different numbers of protons, and their chemical and physical properties are different.
- Isomer atoms are atoms with the same number of neutrons and protons but different energies (Maher, 2011).

3. Types of Radiation

The energy emitted by radioactive nuclei in the form of particles and electromagnetic waves due to their tendency to transform into a stable structure is called radiation. The concept of radiation, previously used for electromagnetic waves, was expanded in 1985 with the discovery of X-rays, radioactivity and electrons (Podgorsak, 2005). Radiation covers a broad spectrum, including all atomic and subatomic particles, microwaves, radio waves, visible light, infrared, X-rays and ultraviolet rays. Radiation, also defined as the emission of energy in the form of electromagnetic waves or particles, is used to refer to rays such as alpha, beta and gamma emitted by radioactive substances. Substances that decompose by emitting alpha, beta and gamma rays to their surroundings are called radioactive substances and have an unstable structure. Depending on their properties, they are classified as ionising or non-ionising.

3.1. Ionising Radiation

Ionisation occurs when an electron is released from an atomic orbit. Losing an electron from its orbit creates a positively charged ion, and adding an electron creates a negatively charged ion. Alpha particles, beta particles, cosmic rays, gamma and X-rays, free neutrons, atoms and artificial radioactive substances cannot be seen by the eye or detected by any sensory organ. They can be measured with special instruments such as Geiger counters. Although they are used in various fields such as medicine, industry, research and construction, they can cause health problems if not used properly. Depending on the dose, exposure of tissues to ionising radiation can cause skin burns, death, cancer, tumours and genetic damage. Therefore, unless adequate precautions are taken to protect against ionising radiation, it poses a threat to all living things (Moseley et al., 1964).

3.1.1. Alpha Decay

In an experiment conducted by Ernest Rutherford in 1903, it was seen that the nucleus of the helium atom was an alpha particle. In other words, the alpha particle is a helium (He) nucleus consisting of two protons and two neutrons. It is positively charged and is denoted by the sign α (alpha) radiation occurs in isotopes with large atomic numbers and generally in naturally radioactive atoms. Alpha particles can be stopped by materials such as paper of very small thicknesses because they have a relatively large electrical charge compared to other types of radiation (Podgorsak, 2005). This charge causes intense ionisation on the path of alpha particles as they pass through matter, causing them to lose their energy more quickly. Accordingly, their reach distances are short, and they do not cause external radiation hazards. The emission of an α particle and the transformation of an element, which is considered heavy and has an atomic number (Z) of 83 and above ($Z \geq 83$), into a different element is called alpha decay (Khan et al., 2014).

3.1.2. Beta (β) Decay

Beta particles, which have high-energy electrons, can have positive or negative charges. It is the transition of the nucleus to a stable state when one of the excess protons or neutrons in the nucleus turns into another (Chang et al., 2014; Khan et al., 2014). This event can occur in three different ways.

3.1.2.1. β-Decay

When excess neutrons are converted into protons in the nucleus, an electron is emitted due to electric charge conservation and is called negatron decay (Podgorsak, 2009).

3.1.2.2. β+ Decay

In this event, which is the conversion of the proton in the nucleus into a neutron, a positively charged electron is also formed, and this is called positron decay.

3.1.2.3. Electron Capture

The atomic nucleus captures the electron in the orbit closest to itself and a proton turns into a neutron (Chang et al., 2014).

3.1.3. Gamma Decay

Gamma rays are emitted when the nucleus, which undergoes one of the α (alpha) or β (beta) decays, cannot reach a stable state and moves to a lower energy state (Chang et al., 2014).

3.2. Non-Ionising Radiation

The definition of non-ionising radiation includes electromagnetic waves, visible light, microwaves and radio waves that are only absorbed when interacting with matter and whose energy is insufficient for ionisation (Shapiro, 1990). Their energy only stimulates a higher energy level and does not produce charged ions. There are different types and therefore different biological effects. High frequencies can cause biological damage similar to that caused by ionising radiation.

3.2.1. Natural Resources

The Earth has a magnetic field, so the compass shows north and south directions. The radially oriented electric field of the atmosphere varies between 50-500V\m (volt/meter) depending on the weather, time of day, season, and latitude (Henriksen et al., 2003).

3.2.2. Man-Made Sources of Different Frequencies

Electromagnetic fields can be generated by the combined use of electric and magnetic fields.

3.2.2.1. Quasi-Static and Static Fields

Examples include magnetic resonance imaging systems used in health care and electric cars, trams and metros used in transport.

3.2.2.2. Low-Frequency Fields

Electric and magnetic fields are assessed separately. These frequencies are used in various technologies.

3.2.2.2.1. Electric Power Lines

The generated electricity is transmitted long distances with high voltage and low current through a grid system, minimising losses. The electric field generated is proportional to the voltage on the line. Ground clearance is increased to reduce the force felt at ground level. The strength of the magnetic field generated varies according to the day, time, season and amount of electricity consumed. The strength of the magnetic field decreases with distance from the line.

3.2.2.2.2. Devices Used in Industry

Different devices used in industry generate fields of different intensities that affect the environment. For this reason, it is necessary to pay attention to the stopping distance according to the frequency used by these devices.

3.2.2.2.3. Home Used Devices

When electrical devices are plugged in, they start producing electric fields. Although it varies depending on the type of device, the amount of effect decreases rapidly as the distance increases. A magnetic field is generated in the device when current passes through it.

3.2.2.3. High-Frequency Fields

They are electromagnetic field radiations, and their frequencies are between 100 kHz (kilohertz) and 300 GHz (gigahertz).

- Radio and television broadcasts: Electromagnetic field strengths are high due to the wide reach of the broadcast. Since AM (amplitude-modulated) transmitters are used over long distances, they are placed at high points where access is difficult due to their high power, and power cuts are applied during maintenance and repair to prevent harm to the personnel working there. FM (frequency modulated) transmitters, on the other hand, have lower power because their reach area is short and therefore, they can be installed in high-rise buildings. Since TV transmitters have the same power, they are similarly mounted on the roofs of buildings.

- Wi-Fi - Wireless Connection Network: It is used to connect or communicate with equipment such as phones and computers to the internet.
- Modem: used in a wired connection.
- Base Stations: Their transmission power is very high. Depending on the position of the station's antennas, it emits high-frequency waves narrowly in the vertical direction but over a wide area in the horizontal direction. Therefore, the radiofrequency area below the station is small, but it increases as it moves away from the station. For this reason, the public should be kept away from these areas, or they should be placed higher.
- Mobile Phone: Although the power level varies depending on the make and model of mobile phone, attention should be paid to the time of use as they emit high frequency in all directions.
- Satellite Dish: Their power is measured in kilowatts and is high. If they are directed to the satellite, their impact on the environment will be reduced.
- Bluetooth devices, baby monitors, microwaves, and radar are other examples.

4. Radiation Sources

Human beings are exposed to natural and artificial radioactive sources throughout their lives.

4.1. Natural Radiation Sources

Radioactive isotopes in the air we breathe, the water we drink, the soil we walk on, the houses we live in, and our bodies make up a significant proportion of the natural radiation we are exposed to. Factors that affect the amount of natural radiation received include local soil structure, building materials, sea level, atmospheric pressure and weather conditions. Cosmic radiation occurs as a result of high-energy protons, helium, electrons, and heavy particles coming to our world from space and the sun. In addition, natural radioactive nuclei such as ^{238} U (Uranium-238), ^{232} Th (Thorium-232), and ^{40} K (Potassium-40), etc. in the earth's crust also cause terrestrial radiation (Podgorsak, 2005). These natural radiation sources are divided into four groups:

1) Cosmic Radiation
2) Terrestrial Radiation
3) Intrabody Irradiation
4) Radon Gas

Radon gas has entered our lives through building materials, cracks, and inadequate ventilation. It accumulates especially in basements. Among the natural resources, the most effective is radioactive gas (Henriksen et al., 2003).

4.2. Artificial Radiation Sources

Most artificial radiation sources are isotopes such as ^{131} I (iodine-131), ^{99m} Tc (Technetium-99m), ^{192} Ir (Iridium 192), and ^{60} Co (Cobalt-60), which are used in diagnosis and treatment. In places where ionising radiation is used, patients as well as employees are exposed to this radiation. Other sources of artificial radiation include equipment such as x-ray machines, fluorescent lamps, smoke detectors, televisions and cathode ray tubes. Reasons affecting the increase in environmental radioactivity including nuclear tests, nuclear fuel supply, radioisotope production, nuclear weapons, and nuclear accidents (Podgorsak, 2005).

5. Radioactivity

Reactions taking place in the nucleus of the atom provide nuclear energy. The change in the quality or properties of the atomic nucleus by bombarding it with high-energy particles is called a nuclear reaction. This bombardment can be with alpha particles, gamma ray photons, neutrons, protons, deuterons, and heavy ions. The nuclear reaction occurs when the target heavy nucleus and the bombarding light particle react. This particle must have sufficient energy to approach the nucleus. After the reaction, a lighter particle and a heavier nucleus are formed.

Atoms can have stable or unstable nuclei. Stability is understood by the ratio of the number of neutrons (n) and protons (p) in the nucleus (n/p). Nuclear stability can be expressed as the resistance of the nucleus to nuclear decay and disintegration (Henriksen et al., 2003). In the equilibrium state in the nucleus, protons and neutrons cannot move away from the nucleus due to their tight bonds with each other. While the n/p ratio is 1 in light elements, the ratio can increase up to 1.5 in heavy elements. As the number of protons increases, the tendency to become unstable increases. If the nucleus is unstable, it wants to become stable by shedding its excess energy. This phenomenon, called radioactive disintegration (radioactivity), cannot be controlled. Unstable nuclei called radioactive nuclei, break down and become more stable by emitting radiation such as alpha, beta particles, and electromagnetic gamma radiation to their surroundings. This phenomenon is called radioactivity. This concept, called Becquerel rays, was discovered on uranium by the French Physicist Antonio Henry Becquerel in 1896, and in 1898, Marie and Pierre Curie named it radioactivity and detected polonium and radium, two hitherto unknown elements that were more radioactive than uranium. There are two types of radioactivity. These are called natural and artificial radioactivity (Podgorsak, 2005).

5.1. Natural Radioactivity

Natural radioactivity is the phenomenon whereby an atom with an unstable nucleus undergoes spontaneous decay into other nuclei, independent of any external influence. Naturally radioactive elements are found in the region of the periodic table with atomic numbers between 81 and 92. Naturally radioactive elements are divided into groups:

- *Radium Group:* This group starts with uranium 238 and decays into stable lead 206. One of the radioactive decay series in nature is the uranium series. The uranium (U) series has existed for 5 billion years and has only completed its first half-life. This series starts with the ^{238}U (Uranium-238) isotope and has 15 product nuclei. First, the ^{238}U isotope turns into ^{234}Th (Thorium-234) nuclei by emitting alpha particles and turns into ^{234} Pa (Protactinium-234) nuclei by emittingβ-particle. This decay continues until it becomes stable and ends with a ^{206}Pb (Lead-206) nucleus.

- *Actinium Group:* This series begins with uranium 235 and ends with conversion to. There are 15 product nuclei in the series, which starts with the ^{235}U(Uranium-235) isotope and ends with the ^{207}Pb (Lead-207) isotope.
- *Thorium Group:* Thorium starts with 232 and ends with lead (Pb) 208. This group is one of the series with the longest half-life. It starts with the ^{232}Th (Thorium-232) isotope and there are 12 product nuclei. The ^{232}Th isotope tries to become stable by emitting an α particle and decays into the ^{228}Ra (Radium-228) nucleus. It then emits a β-particle and turns into a ^{228}Ac (Actinium-228) nucleus. This process continues and ends with the transformation to the stable ^{210}Pb (Lead-210) nucleus.
- *Neptunium group:* In the neptunium series, one of the artificial radioactive series, neptunium originates from ^{241}Pu, which has a half-life of 14 years. The series starting with the ^{237}Np (Neptunium-237) isotope includes the natural elements thallium and bismuth, and this series also includes elements with very short half-lives (Podgorsak, 2005).

5.2. Artificial Radioactivity

Non-radioactive elements can be obtained by bombarding them with high-speed particles (Khan et al., 2003; Pawlicki et al., 2016). While terms such as decay, radiation and emission are used to describe natural radioactivity, the term bombardment is used in artificial radioactivity, and the most appropriate particle for bombardment is the neutron because it is uncharged. Because it is not repelled by the nucleus, it can easily interact. Artificial radioactive nuclei are created by bombarding stable nuclei or long-lived radioactive nuclei (Henriksen et al., 2003).

6. Radioactive Decay

Radioactive decay was discovered by Ruth Herford and Soddy in 1902 (Maher, 2011). Radioactive decay shows that the change in the number of atoms per unit time ($\Delta N/\Delta t$) is proportional to the number of radioactive atoms present (N). The radioactive decay formula is given as "$\Delta N/\Delta t = - \lambda N$.

Λ" is the proportionality coefficient, which is the decay constant. The minus sign indicates that the number of radioactive atoms present decreases numerically with time (Hendee, 1970; Attix, 2008).

$$T=_0 \exp(-\lambda t)=N_0 e\text{-}\lambda t \quad (1)$$

The well-known exponential for radioactive decay is Equation 1. N_0 is the initial number of radioactive atoms and NT is the number of atoms present at time t. The base of the natural logarithm is e = 2.7182 (Khan et al., 2003).

6.1. Activity

The activity of the radioactive substance means the number of disintegrations per unit of time in Equation 2.

$$A= I\frac{dN}{dt}I = \lambda N \quad (2)$$

In this equation the letter A represents activity is equation 2.

$$A_t=A_0\exp(-\lambda t)= A_0 e^{-\lambda t} \quad (3)$$

Here in Equation 3, At is the remaining activity at time t, while A0 is the activity at the initial activity start time and is equal to λN0. The SI (International System of Units) unit used for activity is the Becquerel. In radiotherapy, the more common unit of activity is the Curie (Ci) (1Ci =3.7x10 10 Bq) (ICRU 33, 1980).

6.2. Half-Life

The half of a radioactive substance is the time during which the activity decreases by half from its initial value. It is expressed as $½ = e^{-\lambda t}$ or $T_{1/2} =$ ln2/λ. The "In2" value is used as approximately 0.693 (Henriksen et al., 2003).

6.3. Average Life

The average lifespan (T_a) is found by dividing the total lifespan of the cores by the number of cores used. It is expressed as $T_a\lambda/N_0 = N_0$ or $T_a = 1/\lambda$.

It is used in permanent brachytherapy implant dose rate calculation.

6.4. Effective Half-Life

The half-life of an isotope excreted from the body is shorter than its physical half-life. The physical half-life is called t_p, the biological half-life is called t_b, and the effective half-life is called t_{eff}. It is expressed as in Equation 4 and 5.

$$1/t_{eff}=1/t_b+1/t_p \tag{4}$$

$$t_{eff}=(t_b*t_p)/(t_b+t_p) \tag{5}$$

7. Radioactive Decay Methods

7.1. Alpha Particle Decay

Radioactive nuclides with high atomic numbers mostly undergo alpha (α) particle decay. The unstable nucleus emits a particle called the nucleus of the helium atom, which consists of two protons and two neutrons. In this decay, the mass number decreases by 4. The conversion of radium to radon can be given as an example in Equation 6.

$${}_{88}{}^{226}Ra \rightarrow {}_{88}{}^{222}Ra + {}_{2}{}^{4}He + 4.87MeV \tag{6}$$

Ra: Radium, He: Helium and MeV: Million Electron Volt.

7.2. Beta Particle Decay

Beta (β) decay is radioactive decay that occurs when the nucleus emits negative or positive electrons. Negatron is split into β- or positron, β+, and is formed during the decay process. It is shown like this in Equation 7 and 8.

$$n \rightarrow p^{+}e^{-} \text{ negatron } (\beta-) \quad (7)$$

$$p \rightarrow n^{+}e^{+} \text{ positron } (\beta+) \quad (8)$$

n: neutron, p: proton, e^{-},e^{+}: antineutrino and υ neutrino, and as a result, nuclei that must gain and lose protons are formed (Glasser et al., 1961).

In negatron emission, the neutron ejects an electron (β-) and antineutrino from the nucleus, turning into a proton and sharing its energy. The reaction equation for negatron or β-emission is shown in Equation 9.

$$_{Z}{}^{A}X \rightarrow \; _{Z+1}{}^{A}Y + {}_{-1}{}^{0}\beta + \Sigma\upsilon + Q \quad (9)$$

Positron emission occurs when a positive electron or positron is emitted. The decomposition reaction is expressed as follows in Equation 10.

$$_{Z}{}^{A}X \rightarrow \; _{Z-1}{}^{A}Y + {}_{1}{}^{0}\beta + \upsilon + Q \quad (10)$$

The positron-electron annihilation event is used in positron emission tomography (PET), where it is detected by detectors and the anatomical region is reconstructed by software using a back reflection algorithm.

7.3. Electron Capture

The capture of an orbiting electron into the nucleus and converting a proton into a neutron are called electron capture and is seen as an alternative to positron decay. The neutron deficiency in the unstable nucleus can be made stable by increasing it by electron capture. This is expressed by Equation 11.

$$_{Z}{}^{A}X + {}_{-1}{}^{0}e \rightarrow \; _{Z-1}{}^{A}Y + \upsilon + Q \quad (11)$$

7.4. Internal Transformation

Excess energy is transferred to one of the orbital electrons and removed from the atom. In other words, the internal photoelectric effect in which the γ-ray escaping from the nucleus interacts with the orbital electron is called internal conversion. In this case, the kinetic energy of the electron is found by subtracting the energy released by the nucleus from the orbital electron binding energy. The ejection of the orbital electron creates a hole, resulting in the formation of an Auger electron or characteristic photon. This is expressed by Equation 12.

$$_Z{}^AX^{+} \rightarrow {}_Z{}^AX + e^{\bar{k}} \quad (12)$$

In the isomeric transition, the nucleus loses excess energy as internal conversion or γ-rays. Here the symbol "γ" represents gamma ray. The radioactive nucleus is not destroyed by gamma radiation alone and its excited state continues for a while. When this excited nucleus is in a metastable state, it becomes an isomer of the final product nucleus with the same atomic and mass number but a different energy state (Khan et al., 2014). This is shown in Equation 13.

$$_Z{}^{Am}X \rightarrow {}_Z{}^AX + \upsilon \quad (13)$$

7.5. Alpha, Proton Reaction

The first nuclear reaction was observed in 1919 when Rutherford bombarded nitrogen gas with alpha (α) particles. Alpha proton's reaction is the reaction in which the α particle interacts with the nucleus, forming a compound nucleus, and disintegrates into a new nucleus by expelling a proton (p). This is formulated with Equation 14. Q is the energy released or absorbed in the reaction.

$$_Z{}^AX + {}^4{}_2He \rightarrow {}^{A+3}{}_{Z+1}Y + {}^1{}_1H + Q \quad (14)$$

7.6. Alpha, Neutron Reaction

The emission of neutrons (n) by the nucleus bombarded by alpha (α) particles is called the α, n reaction. As seen in the example of Frederic and Irene Joliot Curie, aluminium produced positrons through an exponential decay. The artificial radioactivity is exemplified in Equation 15.

$$^{27}_{13}Al + ^{4}_{2}He \rightarrow ^{30}_{15}P + ^{1}_{0}n \quad (15)$$

7.7. Proton Bombardment

The most common one, which occurs when protons are captured by the nucleus by γ-ray radiation, is called the p,γ reaction. Protons produced by positron emitters can form the reactions (p,γ) - (p,n) - (p,d) and (p,α).

7.8. Deuteron Bombardment

It is a combination of a proton and a neutron and occurs when the compound nucleus splits as a result of the emission of neutrons or protons.

7.9. Neutron Bombardment

Neutrons, which have no electrical charge, are effective in penetrating the nucleus and producing nuclear reactions. In addition, they do not need to have high kinetic energy to do so. An example of a neutron capture is shown in Equation 16.

$$^{10}_{5}B + ^{1}_{0}n \rightarrow ^{7}_{3}Li + ^{4}_{2}He \quad (16)$$

The formation of a reaction with fast or slow neutrons depends on the mass difference between the product and the bombardment nucleus (Chang et al., 2014).

7.10. Integrated Photons

The interaction of the high-energy photon with the atomic nucleus allows the emission of one or more nuclei and the formation of a nuclear reaction. As a result, neutron emission occurs from the nucleus.

7.11. Fission

It is obtained by bombarding nuclei with high atomic numbers with neutrons. The nuclei absorbed by the neutron turn into nuclei with low atomic numbers and neutrons. Thermal neutrons are more effective in producing reactions. An example of fission with slow neutrons is shown in Equation 17.

$$^{235}{}_{92}U + {}^{1}{}_{0}n \rightarrow {}^{236}{}_{92}U \rightarrow {}^{141}{}_{56}Ba + {}^{92}{}_{36}Kr + 3{}_{0}{}^{1}n + Q \quad (17)$$

7.12. Fusion

Fusion is the opposite of fission. In other words, nuclei with low mass combine to form a new nucleus. For this reaction to occur, they must be close enough to each other to overcome the Coulomb force (Khan et al2014). The reacting nuclei overcome the potential wall between them and achieve fusion. Equation 18 exemplifies this.

$$^{2}{}_{1}H + {}^{3}{}_{1}H \rightarrow {}^{4}{}_{2}He + {}^{1}{}_{0}n + Q \quad (18)$$

8. X-Ray Generation

Bremsstrahlung X-rays and characteristic X-ray formation mechanisms can explain the formation of X-rays.

8.1. Bremsstrahlung

It occurs after a radioactive collision of a high-energy electron with a nucleus. As the electron passes near the nucleus, it is deflected by the Coulomb force, loses energy, and this energy is emitted as electromagnetic radiation.

All or part of the electron's energy is lost when it interacts with one or more braking radiations. In such cases, the braking radiation energy can be at most the leading electron energy. X-rays are propagated in a forward direction due to the kinetic energy of the electrons. While the electrons hit the target, X-rays are obtained from the other side (Glasser et al., 1961; Morgan, 1944).

The energy loss for the electron is related to the square of the atomic number (Z^2). Therefore, the probability of the occurrence of braking radiation varies with the Z 2 of the target material. In addition, the generation of X-rays is primarily related to the voltage applied to the tube and its atomic number (Glasser et al., 1961).

8.2. Characteristic X-Rays

Electrons interacting with the target while passing close produce characteristic X-rays. Electrons with energy E_0 can interact with an atom in the target and ionising the atom by ejecting an orbital electron from the K, L, or M shells. The original electron returns from the collision with energy E_0-ΔE. ΔE represents the energy transferred to the orbital electron, part of which is used to overcome the binding energy of the electron, while the remainder is used for the ejected electron (Hendee, 1970). If a gap forms in the orbital, an electron in the outer orbit fills that gap and the resulting energy (electromagnetic radiation) is given out. This radiation, determined characteristic radiation, reflects the characteristics of the target atom and its transition layers. The transition involves inner layers such as K, L, M, and N in targets with high atomic numbers. So, since the characteristic radiation emitted is high, it is found in the X-ray part of the electromagnetic spectrum.

The characteristic radiation is emitted as discrete energy. When the electron falls from the L-layer to the K-layer, the emitted energy is $h\nu = EK-EL$, where Ek is the electron binding energy in the K-layer and El is the energy in the L-layer. The threshold energy, or critical absorption energy, is the energy an electron must have to remove an electron from another atom.

X-ray tubes are X-ray producing devices that emit X-rays by striking the metal target with electrons accelerated to high energies. The X-ray tube consists of a tungsten anode, which is separated from the filament and placed in a vacuum glass tube, and a cathode containing the filament. X-rays are produced by applying high voltage to the cathode and anode ends of the circuit. The electrons released accelerate towards the anode and strike the target. The effective interactions are bremsstrahlung and characteristic radiation (Pawlicki et al., 2016; Sabattier et al., 2007).

9. Quality Parameters of X-Ray Beams

X-ray beam quality depends on knowing the spectral distribution.

9.1. Half-Value Thickness

The thickness of the absorber metal required to halve the original intensity in the radiation beam is called half-value thickness (HVL). It refers to the quality of the X-ray spectrum. All beams except gamma rays are represented by their HVL, unlike others gamma ray quality is defined by the core or emitted gamma ray energy (Almond, 2009).

The quality term refers to the entire energy of the beam. The energy spectrum of the beam is directly affected by the change in the atomic number (Z) of the target material, the change in filtration, the change in the peak voltage value (kVp), and the applied waveform. The quantity of the beam is related to the number of photons in the beam. When the number of photons increases, the beam intensity increases. As the HVL increases with the thickness of the filter, the high energy photons in the beam will harden, causing the irradiation speed to decrease. With this information, the thickness of the HVL must be adequate for the beam efficiency to be acceptable (Khan et al., 2014).

9.2. Peak Voltage

The characteristic radiation is emitted as discrete energy. When the electron falls from the L-layer to the K-layer, the emitted energy is $h\nu = EK\text{-}EL$, where EK is the electron binding energy in the K-layer and El is the energy in the L-layer. The threshold energy, or critical absorption energy, is the energy an electron must have to remove an electron from another atom.

X-ray tubes are X-ray producing devices that emit X-rays by striking the metal target with electrons accelerated to high energies. The X-ray tube consists of a tungsten anode, which is separated from the filament and placed in a vacuum glass tube, and a cathode containing the filament. X-rays are produced by applying high voltage to the cathode and anode ends of the circuit. The electrons released accelerate towards the anode and strike the target. The effective interactions are bremsstrahlung and characteristic radiation (Pawlicki et al., 2016; Sabattier et al., 2007).

9.2.1. Direct Measurement

9.2.1.1. Voltage Divider

A voltage divider is a circuit created by connecting a number of resistors in series across the high voltage terminals, creating a total resistance. By gaining access to the tube's high voltage leads, the effective voltage passing through the tube can be measured using a voltage divider.

9.2.1.2. Sphere-Aperture Method

In this old method, the power cable in the X-ray tube is connected to the shiny metal sphere using the cable adapter and brought close to the distance at which a spark can occur between the spheres. At this critical distance, the kVp passing through the X-ray tube can be calculated, taking into account corrections for air density and humidity.

9.2.2. Non-Direct Methods

9.2.2.1. Fluorescence Method

When the energy of the incoming photon in the X-ray tube where an absorbing material is placed in the target area is greater than the binding energy of the electron, the electron is removed from the material by absorption. The gaps formed in the orbits are filled with electrons in the outer orbits to have a stable state. In this event, it emits its fluorescence X-rays due to the orbital energy difference (Khan et al., 2003).

This is defined as the "X-ray fluorescence" or "emission" method and is based on two basic principles:

1) Peak potential (kVp) and photon energy (keV) are numerically equal.
2) K orbital emission (characteristic, fluorescence X-ray generation) occurs when the incident photon energy is equal to or greater than the binding energy of the orbital electron. This is known as the K absorption edge threshold value phenomenon. The kVp scale on the device can be calibrated with materials with different K absorption edges.

The gaps formed in the orbits are filled with electrons in the outer orbits to obtain a stable state. In this event, it emits its fluorescence X-rays due to the orbital energy difference (Khan et al., 2003).

This is defined as the “X-ray fluorescence” or “emission” method and is based on two basic principles:

1) Peak potential (kVp) and photon energy (keV) are numerically equal.
2) K orbital emission (characteristic, fluorescence X-ray generation) occurs when the incident photon energy is equal to or greater than the binding energy of the orbital electron. This is known as the K absorption edge threshold value phenomenon. The kVp scale on the device can be calibrated with materials with different K absorption edges.

9.2.2.2. Decrease Method

A detector assembly, such as an ion chamber, consisting of two aluminum or copper containers of different thicknesses is used. Primarily, it is placed in the X-ray beam produced with a known peak voltage measured with two containers and calibrated by proportioning. The permeability curve from high filtration. This method is less sensitive because it is affected by fluctuations in the X-ray tube potential.

9.2.2.3. Penetrometer

It is the comparison of the transmittances of two materials showing different X-ray absorption depending on the photon energy. The device, developed by Benoist in 1901, was revised by Stanton. If the device is calibrated with a known potential, the peak voltage can be read from the calibration curve.

9.3. Effective Energy

The effective energy is the energy equivalent to the photon energy of monoenergetic rays at the same absorption rate. It changes due to the hardening of the beam.

9.4. Average Energy

The spectral distribution of radiation beams is expressed in terms of energy and the distribution of photons or energy flux.

10. Interaction of Ionising Radiation with Matter

10.1. Interaction with Photons

Interaction with matter occurs through photodisintegration, coherent scattering, photoelectric effect, Compton scattering, and pair formation.

10.1.1. Photodisintegration

At photon energies greater than 10 MeV, it occurs between the atomic nucleus and the photon. This sometimes leads to the emission of nucleons and in most cases neutrons.

10.1.2. Coherent Scattering

Rayleigh scattering (coherent scattering) occurs between low-energy photons and substances with high atomic numbers. The electron vibrating with the magnetic wave passing nearby emits waves at the same frequency as the electromagnetic (EM) wave. The photon is scattered at a certain angle, however the energy in the medium is not absorbed and does not change (Khan et al., 2003).

10.1.3. Photoelectric Effect

The interaction seen in low-energy photons absorbed by the atoms of the medium pass through is the photoelectric effect. By absorbing X-rays, mono energetic electrons are produced; these are called Auger electrons. The photoelectric attenuation coefficient (τ/ρ) varies directly proportional to the atomic number (Z^3) of the absorber material and inversely proportional to the energy of the incident photon (E^3). This situation is shown in Equation 19.

$$\tau/\rho \ \alpha \ Z^3 / E^3 \qquad (19)$$

It is the correlation that forms the basis of many applications in radiology. X-ray absorption of tissues is different in bones, fat, and muscles with different atomic numbers. Z^3 is important in imaging with the use of contrast material. In low-energy superficial and medium voltage treatments, the effect of X-ray absorption on bones is related to Z^3.

10.1.4. Compton Scattering

It occurs between photons with energy between 30KeV-30MeV and free or weakly bonded electrons (Attix, 2008). Free electrons are the electrons located in the outer orbit of the atom and have very low binding energy. The incoming photon gives some of its energy to the electron as kinetic energy. Depending on the energy of the incoming photon, the electron and the photon with decreasing energy are scattered at various angles.

10.1.5. Pair Formation

When the incident photon energy is greater than 1.02 MeV, pair production can occur as the photon interacts with matter. The positron produced by pair production loses energy like an electron through ionisation, excitation and bremsstrahlung events. The slowed down positron combines with a free electron to form two photons travelling in opposite directions at an angle of 180° and has an energy of 0.511 MeV. It is the result of interaction with the electromagnetic field of the nucleus and its probability of formation increases with Z.

10.2. Interaction of Charged Particles with Matter

The interaction occurs by ionisation and excitation, and the bremsstrahlung interaction occurs. Scattering occurs more in electrons than in heavily charged particles. Nuclear reactions occur in heavily charged particles through the Coulomb interaction, resulting in radioactive nuclei.

10.2.1. Heavy Charged Particles

More ionisation occurs in heavily charged particles due to their mass and low velocity. The Bragg peak is the dose curve produced by the particle in the narrow region before it stops. It is said that this is where the most ionisation and energy loss occurs. The Bragg peak effect and minimal scattering provide the advantage of high dose in the target volume and minimal dose in the surrounding tissue.

10.2.2. Interaction of Electrons

The interaction of electrons within matter is similar to heavy particles. Due to their small mass, they are under the influence of increasing scattering and their direction of movement changes. As a result, Bragg Peak is not obtained. Its interaction with water or soft tissue occurs mainly in the form of ionisation and excitation, dose accumulation occurs in the environment. If the transferred energy cannot remove the electron from the atom, the electron just moves and then returns, this is called excitation (Khan et al., 2014).

10.2.3. Interaction of Neutrons

Even though neutrons cause indirect ionisation like X and gamma rays, their interactions with matter are different. They interact with matter in two different ways. The first way is with the heavy nuclei of hydrogen, protons, and other elements with Elastic collision. With the collision, energy is redistributed between the particles. The other is nuclear disintegration (Khan et al., 2014). If the masses are similar, energy transfer is efficient; if the collided nucleus is heavy, it loses less energy. For example, if X-rays above 10 MV are used in radiotherapy, the beam interacts with the metal compounds in the device head. It causes photodisintegration and the scattered neutrons can reach the patient.

11. Radiation Dose Units

After the discovery of X-rays by the German physicist Wilhelm Conrad Röntgen in 1895, the unit of radiation quantity was named "Roentgen." The International Commission on Radiological Units and Measurements (ICRU) adopted in 1956 the unit of radiation quantity as roentgen and the unit of absorbed dose as rad (ICRU 33, 1980; ICRU 10b, 1964).

11.1. Exposure Dose Unit

It is the amount of ionisation created by X-ray or γ ray in air. X-ray (R) was used as the first irradiation unit.

X-ray (R): It is the exposure dose unit of the new unit that was accepted as C/kg.

C/kg: It is the amount of x or gamma radiation of 1 Coulomb of positively or negatively charged ions in 1kg of air mass.

11.2. Absorbed Dose

It is the amount of energy stored in a unit mass. Its old usage is Rad.

Rad: It is defined as the amount of 100 erg of energy absorbed by 1 gram of tissue. It is shown in Equation 20.

$$1\text{rad}=10^{-2}\ \text{J/kg}=\text{Gray (Gy)} \quad (20)$$

Kerma: It is the average energy transferred from ambient ionising radiation to charged particles (electrons) per unit mass (dm) and its unit is kilogram per joule (J/kg). It is shown in Equation 21 and 22. The Kerma unit is Gray (Gy).

$$1\ \text{Gy} = 1\text{J/kg (Joule/Kilogram)} \quad (21)$$

$$1\text{rad}=0.01\ \text{J/kg}=100\text{erg/gr} \quad (22)$$

11.3. Equivalent Dose Unit

Equivalent dose unit is a measurement that includes physical and biological factors. It is found by multiplying the absorbed dose in a tissue by the quality factor and other correction factors. Quality factors vary between 1 and 20 depending on the type of radiation and the unit is REM.

Rem: Rem is defined as the radiation dose expressing the biological effect caused by 1 Roentgen portion of X or gamma radiation. It is used as Sievert in the MKS unit system. It is explained in Equation 23.

$$\text{Equivalent dose} = \text{physical dose x quality factor} \quad (23)$$

If we need to provide examples of quality factors according to the type of radiation;

- The quality factor is 1 for x-rays, gamma rays, electrons, and beta rays with maximum energies greater than 0.03MeV.
- For beta rays with maximum energies not greater than 0.03 MeV, 1.7 is used.
- It is 0 for naturally occurring alpha particles and 20 for heavy recoil nuclei.

Sievert: The biological effects caused by different ionising radiations are different even if the absorbed energy is the same. As LET increases, the biological effect also increases. The equivalent dose unit is expressed as J/kg in an SI unit, its name is Sievert (Sv) that is the biological dose unit used for the human body, organs, and tissues. It is expressed as the radiation dose corresponding to the energy absorption of 1J per 1kg of the tissue or organ. It is shown in Equation 24.

$$1Sv\ (Sievert) = 1J/kg = 100rem \tag{24}$$

LET: An electrically charged particle travelling through matter loses energy by interacting with atoms along the way, and this energy is transferred to the atoms. The particle loses energy as it travels. Linear energy transfer (LET) is usually measured in kiloelectron volts per micron, KeV/μ. The LET values of particle radiation are large. Radiobiological effect values depend not only on the type of radiation, but also on the biological effect received. A particle with a high LET value also has a high RBE value.

11.4. Activity

Activity expresses the number of disintegrations of a radioactive isotope in unit time and its unit is Curie (Ci). The activity is measured in seconds. The amount of activity of 1 gram of Ra-226 is called 1 Ci. On the other hand, the new unit Becquerel is used as the special unit of activity and one nuclear fragmentation per second is called "1 Becquerel." In 1985, ICRU proposed to define radiation measurement units in the MKS unit system and new units were subsequently used.

Conclusion

Radiation is energy that comes from a source and is carried in the medium by particles and electromagnetic waves (mass less packets of energy called 'photons'). The transfer of energy to an atom and the removal of electrons from the atom is called ionisation. Radioactive decay is the process by which an unstable atomic nucleus loses energy through radiation. A material containing unstable nuclei is called radioactive. Unstable nuclei have different decay characteristics depending on their nuclear properties. The type of decay depends on the number of protons and neutrons in the nucleus and the distance from the stability region. It may take a single decay or several decays for radioactive material to reach a stable state. The ionising radiation produced by these decays causes various physical, chemical and biological changes in living organisms at the molecular and cellular levels. The biological effects in the cell depend on the radiation dose, the dose transmission rate, the type of radiation, the energy of the radiation, the distribution of the dose in the tissues and the sensitivity of the tissues to radiation.

References

Almond, P. R. (2009). A historical perspective: a brief history of dosimetry, calibration protocols, and the need for accuracy. *Clinical Dosimetry Measurements in Radiotherapy,* edited by DWO Rogers and JE Cygler, 1-27.

Attix, F. H. (2008). *Introduction to radiological physics and radiation dosimetry*. John Wiley & Sons.

Chang, D. S., Lasley, F. D., Das, I. J., Mendonca, M. S. and Dynlacht, J. R. (2014). *Fundamental physics and biology of radiotherapy* (No. 14711). Springer International Publishing.

Glasser O., Quimby E. H., Taylor L. S., Weatherwax J. L. (1961). *Physical Foundations of Radiology.* 3rd ed. New York: Paul B. Hoeber. Medical book Department of Harper & brothers.

Hendee, W. R. (1970). *Medical Radiation Physics.* United Kingdom: Year Book Medical Publishers.

Henriksen T., Maillie, H. D., (2003), *Radiation and Health,* ISBN 0-415-27162-2.

International Commission on Radiation Units and Measurements. (1980*). Radiation Quantities and Units.* Report No. 33. Washington, DC: International Commission on Radiation Units and Measurements.

International Commission on Radiation Units and Measurements (1964). *Physical Aspects of Irradiation.* Report No. 10b. *Handbook 85.* Washington, DC: NBS.

Kamal A. (2014). *Nuclear Physics.* Murphy, TX, USA.

Khan F. M., Gibbons J. P., Mihailidis D. N., Aikhatib H. (2003). *Handbook of the Physics of Radiation Therapy.* Philadelphia: USA.

Khan F. M., Gibbons J. P. Nuclear Transformations. (2014). *Khan's Physics of Radiation Therapy.* 5th ed. USA: *Lippincott Williams and Wilkins.*

Maher K. (2011). *Basic Physics of Nuclear Medicine.* Platypus Global Media.

Morgan, R. H. (1944). A simple method for measuring peak voltage in diagnostic roentgen equipment. *Am. J. Roentgen. and Radium Ther., 52,* 208-211.

Moseley Jr, R. D., & Rust, J. H. (1964). *The reduction of patient dose by diagnostic radiologic instrumentation.* (No Title).

Podgorsak E. B. (2009). *Kinetics of Radioactive Decay, Modes of Radioactive Decay. Radiation Physics for Medical Physicists.* 2nd ed. Heidelberg Dordrecht, London, New York: *Springer.*

Podgorsak E. B. (2005). *Basic Radiation Physics. Radiation Oncology Physics: A Handbook for Teachers and Students.* 1st ed. Vienna: International Atomic Energy Agency.

Pawlicki, Todd, Scanderbeg, Daniel J., Starkschall, George. (2016). *Hendee's Radiation Therapy Physics.* 5th ed. Hoboken, New Jersey: John Wiley& Sons, Inc.

Sabattier, R., Jakel, O., & Mazal, A. (2007). High-LET modalities. Mayles, P., Nahum, AE, and Rosenwald, J.-CI. (Eds.), *Handbook of Radiotherapy Physics: Theory and Practice,* Taylor & Francis, New York, 1053-1068.

Shapiro J. (1990). *Radiation Protection: A Guide for Scientists and Physicians,* 3rd ed., Cambridge: Harvard University Press.

Unscear, S. (2000). *Effects of Ionizing Radiation.* United Nations, New York, 453-487.

Chapter 2

Basic Principles of Radiation Protection and Radiation Protection Methods

Nural Öztürk*

Department of Radiation Oncology, Aydın Adnan Menderes University, Aydın, Türkiye

Abstract

Electromagnetic waves, or subatomic particles, have energy that is defined as radiation, which has natural and artificial types. Although it is not possible to be protected from natural radiation, for artificial radiation just the opposite is valid. It is important to apply the criteria recommended by international committees. The principles of radiation protection are defined as triplicates, justification, optimisation, and dose limits. The optimisation principle is also defined as Low As Reasonably Achievable (ALARA) and includes three rules time, distance, and armouring. Apart from this, recent international organisations have updated the radiation and tissue weight factors in their latest reports, based on the latest studies examining the radiobiological consequences of radiation exposure. In addition, these updated reports also include radiation damages depending on equivalent and effective dose amounts. Over the years, many reports proposed by committees around the world contain recommendations for the implementation of basic radiation protection principles. Epidemiological studies concerning Japanese survivors of the atomic bomb explosions in Hiroshima and Nagasaki are supported by data from individuals exposed to radiation during medical practice. The data include individuals exposed to radiation through their occupations, the public exposed to high levels of natural radiation, and animal experiments on radiation from individuals exposed to high

* Corresponding Author's Email: nural.ozturk@adu.edu.tr; nuralo@gmail.com.

In: Triage in Radiation Accidents
Editors: Nural Öztürk and Sibel Karaca
ISBN: 979-8-89530-415-0

levels of natural radiation. In addition, these studies also address radiation safety practices in industrial and medical institutions, the control of radionuclides in the environment, protection of the public, and the evaluation of radiation risk. From these reports, according to ICRP 103, emergencies emphasise that in cases of exposure during work, first protection strategies should be implemented and then leakage doses should be taken into consideration. Although current radiation protection standards are generally considered to be acceptably robust, there is significant scientific uncertainty regarding health risks, especially at low doses or low dose rates. Awareness of this issue tends to increase around the world. This chapter examines the basic rules for protection from ionising radiation under international protocols. It also highlights what needs to be done at the national and international levels to prevent exposure to ionising radiation, including for future generations.

Keywords: ALARA, application of protection principles, basic principles, radiation protection principles

1. Introduction

Radiation is defined as energy in the form of electromagnetic waves or subatomic particles, which can be described as the spontaneous change due to the emission of unstable atomic nuclei. Since the discovery of the X-ray by Wilhelm Conrad Roentgen in 1895, many uses of ionising radiation and radioactivity have been discovered and have become widely used, especially in the diagnosis and treatment of diseases (Farmelo, 1995). The penetrating property of ionising radiation is used for health. In light of technological developments, the use of X-rays has an important place in human studies and the treatment of diseases. As a result of the development and testing of nuclear weapons after the Second World War, it was shown that ionising radiation could cause harm as well as its benefits and that people should be protected from radiation while working with ionising radiation (Wojcik & Harms-Ringdahl, 2019).

It is impossible to protect against ionising radiation emitted from natural radiation sources. Furthermore, the results of the nuclear bombs dropped on Hiroshima and Nagasaki, the nuclear tests carried out in various parts of the world, and finally, the Chornobyl and Fukushima reactor accidents have been investigated, and it has been concluded that ionising radiation is harmful, its chronic effects cause serious consequences, and that its exposure can continue even after years (Sebastian Pflugbeil, 2011; UNSCEAR, 2016). Decisions regarding radiation protection standards around the world are based on scientific information updated and reviewed by the United Nations Scientific Committee on the Effects of Atomic Radiation (UNSCEAR) (UNSCEAR, 1988), the International Commission of Radiation Protection (ICRP) (ICRP 103, 2007) and it depends on the publications and bulletins of the International Atomic Energy Agency (IAEA) (IAEA, 2023). Various similar independent organisations and National Committees are formed using the work of these organisations.

Among these organizations are the World Health Organization (WHO), the Nuclear Energy Agency of the OECD (NEA), the International Labor Organization (ILO), the Food and Agriculture Organization of the United Nations (FAO), and the Pan American Health Organization (PAHO). It is also available from the National Council on Radiation Protection and Measurements (NCRP), which reports current recommendations regarding ionising radiation based on radiation protection principles that evolve and change depending on the effects of radiation on human populations.

Another goal of the NCRP is to cooperate with other national and international, public and private organizations regarding radiation quantities, units, and measurements, especially with the ICRP regarding radiation protection. Therefore, NCRP's basic principles and recommendations for radiation protection are closely related to ICRP reports. Thus, as mentioned earlier, the basic principles of the radiation protection system (justification, optimisation, dose limits), initially stated in ICRP Publication 26 (ICRP 26, 1979), were adopted and implemented in the recommendations of Report-91 (NCRP 91, 1987). These principles are based on recent research in ICRP Publication 60 (Smith H., 2016) and NCRP Report No. 116 (Charles B. Meinhold, 1993; Kase, 2004). These committees around the world have been publishing recommendations regarding protection from ionising radiation for more than fifty years. Their publications shed light on efforts to reduce patient radiation doses to prevent unnecessary radiation exposure in a large part of the world. The studies are not finished, they are still ongoing (Mettler et al., 2020; Radiation Protection N° 180, 2015; Rehani et al., 2020; Kachelrieß & Rehani, 2020). The use of rules and dose limits produced based on new scientific information obtained as a result of the research is important in improving standards for the public, radiation workers, and patients in terms of radiation protection (Salomaa et al., 2017).

2. Radiation Protection Systems and Organizer Principles

Radiation protection aims to protect people and living beings against the potentially harmful effects of radiation while allowing beneficial exposure. The principles of radiation protection systems have been widely applied throughout the world and were first mentioned at the International Congress of Radiology in 1928. Since then, radiation protection systems have been developed by considering the information obtained in all radiation research groups, especially the ICRP organisation, and the possible effects of radiation on plants and animals (flora and fauna).

2.1. International Commission on Radiological Protection (ICRP)

The work of the ICRP provides a convenient way to adopt an appropriate level of protection against the hazardous effects of ionising radiation without restricting the beneficial uses of radiation (Donald A. Cool, 2015). A summary of the protection systems of the ICRP, defined in 2017, which are now up-to-date, is shown in Figure 1.

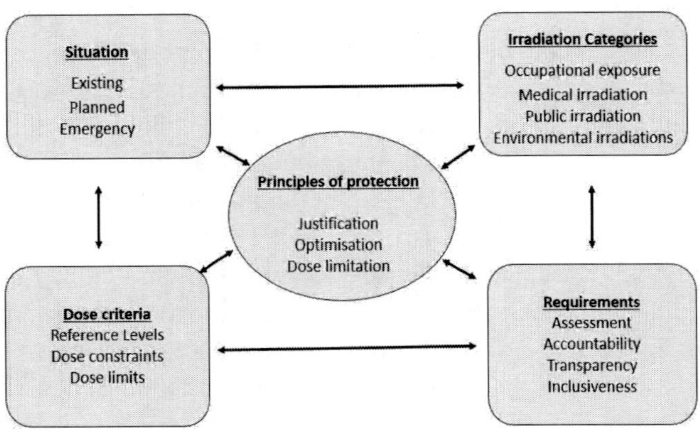

Figure 1. Summary of ICRP's protection systems published in 2017.

ICRP has five main areas of work that are given in the following points:

- Current exposures
- Lessons from emergency accidents and nuclear power plant accidents

- Basics of radiation protection
- Integration of environmental protection
- Support other committees by updating reports.

The ICRP is not an official organization, but a scientific commission that prepares recommendations on radiation protection.

3. Basic Principles in General Radiation Protection

Ionising radiation is used in many areas for the benefit of humanity, especially in health, industry, materials, space, the environment, agriculture, food, and animal husbandry. Without the radiation protection system implementation, there may be harmful effects that affect not only today but also future generations. Therefore, the radiation protection system is very important for protecting future generations from harmful effects and is one of the most effective protection systems used worldwide.

The ICRP-26 (ICRP 26, 1979) report revised in ICRP 60 (Smith H., 2016) presents radiation protection principles, and radiological protection was added to this report. In the report, the protection system is defined separately for '*applications*' and '*interventions*'.

In the report, the radiation protection system has been made even more comprehensive with the three principles of justification, optimisation, and dose limitation.

The main features of these recommendations are given below:

- Updating the concept of effective dose and the radiobiological damages caused by radiation exposure to live tissues;
- Adopting the commission's three basic principles of radiological protection helps define how these principles should be applied to ionising radiation sources and the health effects for individuals exposed to radiation (Smith H., 2016).

After reviewing current biological and epidemiological information on the health risks caused by ionising radiation, it has been reported that the distribution of risks to different organs/tissues, especially in terms of breast cancer and hereditary disease risks, has changed in recent years considering research, and the total damages resulting from cancer and hereditary effects that remain around 5% per Sievert (Sv). The Commission has also hypothesised that the cancer risk from prenatal exposure to ionising radiation would be similar to the cancer risk from early childhood exposure. Considering the results of individual studies in the health field, it is said that computed tomography (CT) scans used for diagnostic purposes account for about 3-7% of all radiological examinations and that CT scans account for 41% of the dose received from diagnostic radiology in some countries (UNSCEAR, 2015; Holmberg et al., 2010).

According to UNSCEAR, it is estimated that approximately half a million patients benefit from CT scans every day. Studies show that global CT use continues to increase, despite the availability of several imaging techniques that do not use X-rays (UNSCEAR, 2000; Schauer & Linton, 2009). Cumulative radiation doses due to frequent or repeated CT scans during diagnosis, especially in emergency department patients who are exposed to these doses are at higher risk for possible carcinogenesis caused by radiation (Griffey & Sodickson, 2009).

In such applications, the application of dose reduction techniques (low milli ampere seconds (mAs), high kilovolts (kV), single shot), and the use of imaging techniques that do not emit ionising radiation (Ultrasound, Magnetic Resonance Imaging) reduce possible cumulative radiation doses, (Gökharman et al., 2016). As a result, it is aimed at reducing the risk of cancer in the population. Considering the new information, the report mentions that a threshold dose must be present for malformations to be induced and severe mental retardation to occur. The report mentions that considering new information, a threshold dose must be present for malformations to be induced and severe mental retardation to occur. The updated version of ICRP-60 used today is also available as ICRP-103 (ICRP 103, 2007). This report retained the effective dose and equivalent dose limits given for skin, hands/feet, and eyes in Report 60, but explained that more information was needed, and revised decisions were required, especially regarding the eye.

The difference between ICRP 60 and ICRP 103 in terms of exposure situations is that response situations are subdivided into 'current' and 'future' exposure situations. Practices have been renamed 'planned' exposures in the report. Intervention levels in terms of individual avoidable doses have been replaced by individual dose reference levels in ICRP 103 (Hedemann-Jensen & McEwan, 2011). As a result of a comprehensive examination of the effects of ionising radiation on health, using the ICRP reports in NCRP Report-116 the recommended dose values that are shown in Table 1, 2, 3. Dosage values in NCRP's advisory guidelines published since 1991 are still valid (Charles B. Meinhold, 1993).

The concepts of effective dose equivalent, effective dose, equivalent dose, and dose equivalent are different (Charles B. Meinhold, 1993). Brief explanations and values of these concepts are summarised in Table 4. (Öztürk N., 2020). The use of equivalent and effective doses from external radiation sources and radionuclides has not changed in the report.

Table 1. Possible estimated risks due to radiation doses to the radiation worker, the public, and the embryo/fetus according to protocols. Sv: Sievert

Group	Risk Factor	ICRP-60	NCRP-91	NCRP-116
Radiation Worker	Terminal Cancer	4% per Sv	1.25% per Sv	4% per Sv
Radiation Worker	Non-Fatal Cancer	0.8% per Sv	-	0.8% per Sv
Radiation Worker	Severe Genetic Effects	0.8% per Sv	0.4% per Sv	0.8% per Sv
People	Terminal Cancer	5% per Sv	-	5% per Sv
People	Non-Fatal Cancer	1% per Sv	-	1% per Sv
People	Severe Genetic Effects	1.3% per Sv	-	1.3% per Sv
Embryo/Fetus	-	-	20% per Sv.	-10% per Sv

Table 2. Radiation worker exposure limits according to different protocols. mSv: milli Sievert

Risk values and affecting factors	ICRP-60	NCRP-91	NCRP-116
Stochastic effects	50 mSv annual effective dose limit mSv at a 5-year cumulative effective dose limit	Annual effective dose equivalent: 50 mSv for the radiation environment 10mSv x age(y) cumulative dose limit	50 mSv annual effective dose limit 10mSv x age(y) cumulative dose limit
Deterministic effects	mSv equivalent dose limit for the lens of the eye mSv annual equivalent dose limit for extremities (hands and forearms, feet and lower legs)	Annual equivalent dose limit: mSv for eye lens Equivalent dose limit for other organs: 500 mSv (e.g., bone marrow, breast, lung, ovaries, skin, extremities)	mSv annual equivalent dose limit for the lens of the eye 500 mSv annual equivalent dose limit

Table 3. Public exposure limits according to different protocols. ALI: Annual Limits of Intake, mSv: milli Sievert

Risk values and affecting factors	ICRP-60	NCRP-91	NCRP-116
Stochastic effects	Effective Dose equivalent limit: 1mSv per year If necessary, a higher dose should be administered, provided that the average of 5 years does not exceed 1 mSv.	Effective Dose equivalent limit: 1mSv per year Effective Dose equivalent limit: Rare 5mSv	Effective Dose equivalent limit: 1mSv per year Effective Dose equivalent limit: Rare 5mSv
Deterministic effects	Annual equivalent dose limit: lens of the eye 15 mSv 50 mSv for skin, hands and extremities	50 mSv for eye lenses, skin and extremities	50 mSv lens of the eye, skin, and extremities
Embryo/fetus	Dose equivalent limit: 2 mSv to the woman's abdomen from when the pregnancy is known. *Radionuclide uptake:* 1/20 of ALI	Dose equivalent limit: 0.5 mSv per month from when pregnancy is known	Dose equivalent limit: 0.5 mSv per month from when pregnancy is known
Negligible individual Dose (NID)	-	Annual dose equivalent limit: 0.01 mSv per application	An effective dose of 0.01 mSv per year per application.

Table 4. Dosage units and concepts

	Explanation	Equivalent
rem (x-ray equivalent - tissue)	1 Roentgen is defined as any radiation dose that produces the biological effect of X or gamma radiation. Different types of radiation have different biological effects. A quality factor is defined for each type of radiation to calculate the biological effect of radiation on tissue. The rem equivalent dose is equal to the dose in rad multiplied by the quality factor (Q). For x and gamma radiation, the quality factor is taken as one, i.e., equal to rad. For alpha radiation, the quality factor is taken as 10, i.e., rem is equal to 10 times rad. For protons and neutrons up to 10 MeV, Q is usually taken as 10.	rem=rad x Q
Sievert (Sv)	Equivalent unit equal to 100 rem.	1Sv=100rem Sv = Gy x Q
rad (absorbed radiation dose unit)	Rad is a measure of the amount of energy absorbed per gram of tissue in 100 ergs.	1rad=100erg/gram

	Explanation	Equivalent
Gray (Gy) (absorbed radiation dose unit of)	Gray is a measure of the amount of energy absorbed per kilogram of tissue.	1Gy=100rad 1Rad=0,01Gy= 1cGy
Curie (Ci)	Radioactivity, the conventional unit equal to one gram of pure radium-226 radioactivity. The amount of any radioactive isotope that produces $3.7.10^{10}$ decays per second.	1Ci= $3.7.10^{10}$ Bq
Becquerel (Bq)	A standard international unit of radioactivity is the amount of any radioactive isotope that produces 1 decay per second.	1Bq=2.763. 10^{-11} Ci=27 pCi
Per second fragmentation speed (dps)	Atom from the core per second reveals coming out atom six particles number (e.g., alpha particles) or photons (gamma rays). One dps = 60 dpm (per minute fragmentation speed).	1dps=1Bq

As a result of reviewing the available data on the biological effect of different radiations, the weighting factors have changed for radiations with high Linear Energy Transfer (LET), such as neutrons and protons; neutron values are given as a continuous function (Charles B. Meinhold, 1993).

The change is that instead of various mathematical models, doses coming from external and internal sources will be calculated using random phantoms like the human body.

4. General Radiation Safety Rules

It is impossible to completely avoid ionising radiation risk exposure, due to the continuous existence of natural radiation in the environment. As a result, the human body receives a dose that corresponds to half of the annual human radiation dose. The remaining part originates from the ionising radiation used for disease diagnosis and treatment.

It should not be forgotten during exposure to radiation that heard, tasted, smelled, or felt by contact cannot be seen. The existence of radiation is possible only with radiation-measuring devices suitable for measuring possible radiation (Sowby, 1981a; IAEA BSS 115, 1996). Therefore, just as in external radiation, three basic rules must be followed to protect against radiation in devices that emit and produce ionising radiation (Jae Hun Kim, 2018).

4.1. The Rule of Time

Since the radiation dose exposed at a point is proportional to the time spent at that point, every procedure related to radiation must be designed in advance and the procedures to be performed must be determined and done quickly without rushing (equation 1.). At the same time, training and experience in the operations performed will contribute greatly to reducing this time (Figure 3 a).

$$\textit{Received Dose} = (\textit{Dose rate}) \times (\textit{Time}) \qquad (1)$$

For example, if the measuring device shows the presence of a radiation source of 20 mSv/hour in the environment, then in this environment the following points gain importance;

Received dose after 1 hour = (20mSv/hour) x (1hour) = 20mSv

When 2 hours are left, then the received dose = (20mSv/hour) x (2 hours) = 40 mSv.

4.2. The Distance Rule

As one moves away from the radiation source, the radiation intensity decreases with the square of the distance (Inverse Square Law). (Figure 2, Figure 3b) (Sowby, 1981a).

A decrease in the radiation intensity means a decrease in the exposed radiation dose. Therefore, any procedure that requires the use of radiation should be performed as far away from the radiation source as possible. We can calculate the dose using the equation in the following equation 2.

$$(I_1)/(I_2) = (d^2_2) / (d^2_1) \qquad (2)$$

In this equation, I shows the dose rate and d indicates the distance of how far away from the radiation source (Figure 2). When working with radiation, a safe distance should be maintained from the radioactive source.

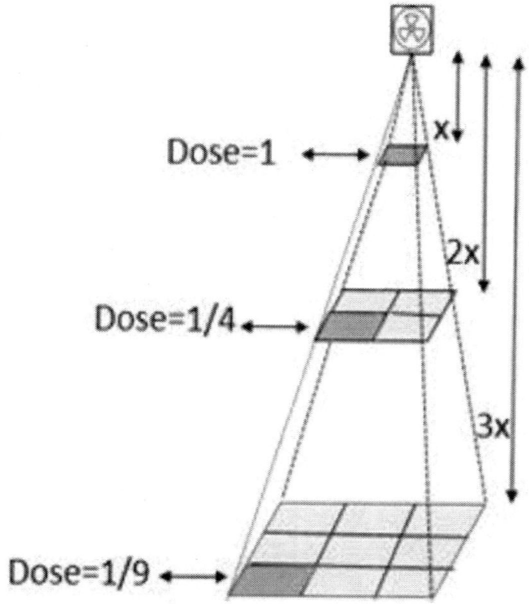

Figure 2. Inverse Square Law. During operation, when the distance from the source is doubled under operating conditions, the dose rate decreases to ¼ of the original value. Similarly, when the distance from the source is halved, the dose rate increases to 4 times the original value.

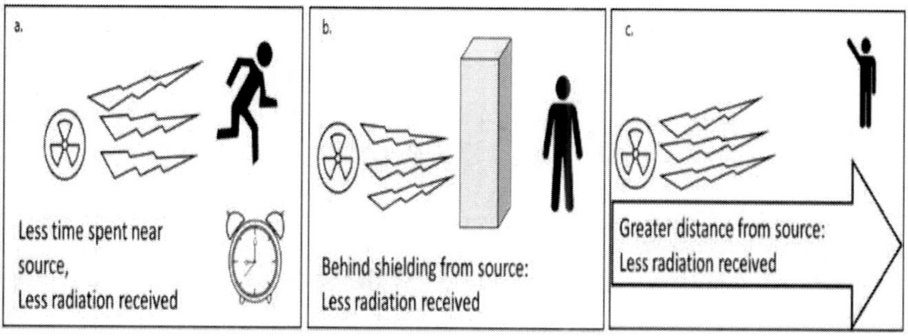

Figure 3. a) Time and exposure are directly proportional. The less time spent in the radiation environment, the less dose is received. b) As one moves away from the radiation source, the radiation intensity decreases with the square of the distance (Inverse Square Law). c) Protective armour such as movable screens, lead aprons, lead-equivalent glass, and screens appropriate to the type and/or energy levels of radiation being studied should be used.

4.3. The Shielding Rule

In cases where parameters such as time and distance cannot be met, radiation-blocking protectors such as concrete walls and lead barriers should be placed between the source and the applicator to reduce the radiation dose received (Figure 3c.) (Hedemann-Jensen & McEwan, 2011).

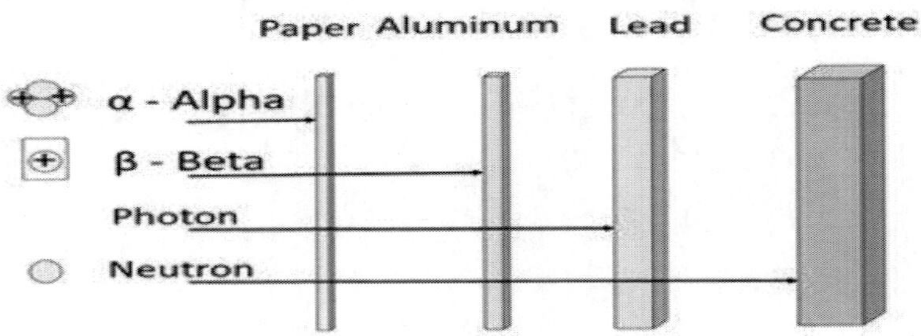

Figure 4. Shielding applied according to radiation type.

Whatever type of radiation one is working with, protective equipment such as moving screens should be used as lead aprons, lead equivalent glass, and screens suitable for the energy levels of that radiation. If one is working with high-energy radiation, then shielding operations must be calculated and carried out meticulously (Figure 4.) (Khan, 2003).

Armouring reduces exposure: Armouring can be calculated using the following expression equation 3.

$$\mathbf{I = I_0 * e^{(-\mu x)}} \quad (3)$$

Here

μ- linear attenuation coefficient of the protective material (1/cm)
x – Thickness of the protective material (cm)
I – Initial exposure rate (intensity of incoming photon)
I – Exposure rate after transmission through the protective material

When a large protection thickness is used in shielding (high Z materials, e.g., lead, steel, concrete, etc.) - it reduces the exposure to gamma/X-ray radiation. Such protective equipment should be used only in areas where low-energy radiology devices are used and in the application room during procedures (Khan, 2003).

When using devices that produce and emit ionising radiation for diagnosis and treatment purposes, one can greatly reduce the dose of radiation he will be exposed to by keeping the irradiation time as short as possible, staying at a far distance from the radiation source during the procedure, and using the necessary protective equipment according to the energy level of the radiation (ICRP 60). For this reason, in this regard, it is important for personnel working with radiation to know what energy level of radiation they are working with and to be aware of self-protection. For this purpose, it is necessary to provide annual training on this subject to personnel working with radiation and to raise awareness about radiation protection. In addition, international reference levels should be followed when diagnosing and adjusting the treatment dose to be applied to patients who will undergo the procedure. Before any type of radiation application, patients and their relatives, including the patient should be informed about the examination and treatment to be performed concerning the radiation dose that may be received, and its risks and benefits.

However, in recent years, with the development of technology resolution images can be obtained at low doses, especially in radiology. Depending on the suitability of the developing technology and treatment techniques in radiotherapy, high doses are given to the tumour, while low doses to surrounding critical organs. This ensures that the doses to be received for both the staff and the patient are low in terms of radiation safety.

5. Basic Principles of Radiation Protection

ICRP has detailed the three basic principles of radiological protection (Justification, Optimisation, and Dose limits) in its latest reports (J. Valentin, 2007). While the principles of Justification and Optimisation are valid in the case of all three basic principles, the application of the defined dose limits is valid in the case of exposure to ionising radiation. Detailed explanations are as follows.

5.1. Principle of Justification

Any application that does not provide a clear benefit due to the harm of radiation should not be allowed. This statement means that to use a radiation source, the possibility of exposure or the reduction of radiation exposure, if any, must be obtained by the societal or individual benefit to reduce the harm caused by this exposure. In these cases, when making decisions, not only scientific opinions but also social, economic, and ethical factors should be considered. The report states: "The scientific committee can make an assessment and provide information about the risks, but ultimately it is the society that will decide on the justification of the risk-cause application through the operation of democratic processes."

The principles should be applied on a case-by-case basis. People who decide to apply radiation should determine the reasons for the application.

For any medical intervention or procedure requiring the use of ionising radiation for diagnosis and treatment purposes to be performed, it must be shown that the benefits to be obtained from them outweigh the risks that may arise from their use. On the other hand, if the same diagnosis or treatment can be performed without the use of ionising radiation, non-ionising radiation should be preferred (IAEA BSS 115, 1996; Sowby, 1981b; Do, 2016). Healthcare personnel must consider the benefits of each irradiation before applying X-rays. In addition, the potential increase in cancer formation should be evaluated against the expected benefit of a definitive diagnosis.

The ICRP commission divides exposure into three different categories *occupational exposure, public exposure*, and *medical exposure* of patients (and companions, caregivers, and research volunteers) to radiation (*healthcare exposure*).

5.1.1. Occupational Exposure

It is defined as all radiation exposure that employees are exposed to as a result of their work. This incident is considered the employer's responsibility and the radiation exposure at the workplace is classified as 'occupational exposure'. The main responsibility for protecting radiation workers belongs to the workplace (Do, 2016).

In the reports, it is stated that if a female employee working in a radiation environment declares that she is pregnant, additional controls should be considered to obtain a level of protection for the embryo/fetus with dose limits like the level of protection provided for the public. Even if additional controls are provided, in cases where there is a possibility of a radiation accident, the radiation worker's place of duty must be temporarily changed to a radiation-free environment (ICRP 103, 2007; NCRP 91, 1987; Charles B. Meinhold, 1993).

5.1.2. Public Exposure

This includes public exposures of the except patient occupational exposures and ionising radiation applications. It originates from different radiation sources. Public exposure from natural sources is the most important reason, but this does not reduce the need for protection from artificial radiation. According to reports, exposure of a pregnant radiation worker (and therefore the embryo and fetus) to radiation is considered public exposure and is regulated (Sowby, 1981b).

5.1.3. Health Exposure

In healthcare, patient exposure to radiation is intentional. Apart from radiotherapy, the purpose of radiation application is not to give a radiation dose, but to use radiation to diagnose the disease or in interventional applications (biopsy, angiography, etc.). The medical use of radiation for patients benefits individual healing directly with acceptable expectations. Volunteering is essential before implementation. Before administration, varying degrees of informed consent are required from the patient, including the expected benefit and the potential risks. The amount of information for informed consent varies depending on the level of exposure (e.g., diagnostic, interventional, or therapeutic) and the potential medical emergencies that may be associated with radiation exposure. If one adapts the radiation protection principles, which are the aforementioned standard three basic rules for medical fields from ICRP Recommendation (J. Valentin, 2007).

5.1.3.1. Principle of Optimisation of Protection

The probability of exposure, exposed persons number, and the individual dose amount must be as low as achievable based on economic and social factors. Screening is performed to identify a disease process that has not manifested clinically. Current screening practices using ionising radiation are recommended for certain populations. Patients who undergo screening should be fully informed of the potential benefits and whether individual or societal benefits should be achieved as a result of the application.

5.1.3.2. Principle of Optimisation of Dose Limits

Other than medical exposure to patients, total dose administration should not transgress the recommended limits. In biomedical research, voluntary exposure generally represents social and individual benefits based on the required information content. The three basic medical recommendations for radiation use require special guidance (Zeyrek, 2013).

5.1.3.3. Diagnostic Reference Levels

The diagnostic reference level (DRL) is valid for medical radiation applications. They contribute to radiological applications in medicine and their numerical values are among the recommendations. The values selected may be specific to a country or region. DRL can be used to improve the distribution of observed results across regional, national, or local areas for general medical imaging (Zeyrek, 2013).

5.2. Optimisation Principle (ALARA)

The principle of optimisation of protection applies only if there is a justification decision. During application, all irradiations should be kept at the lowest possible dose value (ALARA-as low as reasonably achievable) (Figure 3, Figure 5) (IAEA BSS 115, 1996; Sowby, 1981b).

The purpose of optimisation (or ALARA) is not to reduce irradiation to zero but to reduce the risks to an acceptable level under the circumstances. During health applications, this optimisation can be achieved by reducing the size of radiation sources, limiting the working time of personnel exposed to irradiation, increasing the distance between people and radiation sources, and using shielding materials. One of the factors to be considered in the optimisation process is the number of people exposed to irradiation in any given procedure and the regional distribution of irradiation doses. According to ICRP reports stated: "the probability of exposure to ionising radiation should be kept as low as possible, taking into account the number of people exposed and the size of their doses, as well as economic and social factors."

This means that an environment in which the dose limits are not exceeded must be satisfied 100%, and at the same time, the radiation doses to which people will be exposed must be kept below these limits and at the lowest possible level with necessary investments (such as increasing armouring, introducing new technology, etc.) to the extent of economic power (IAEA BSS 115, 1996; (Sowby, 1981b). In this application, options that may result in large doses should not be considered at the planning stage. Importantly, these restrictions on doses are applied prospectively, as in the optimisation principle. If it is determined that the reference level is exceeded despite optimised protection strategies, that is, the dose level is reduced, the reasons for this must be investigated.

5.3. Dosage Limitations

In cases of planned exposure of patients other than medical irradiation, it is necessary to pay attention to the total dose limits given to any individual from the sources. In these applications, the dose limits should not be exceeded as recommended by the commission. This principle was introduced to control the stochastic effects of radiation, such as cancer and hereditary damage, on individuals and future generations. It requires a strict limitation of the effective equivalent dose that a person can receive annually (Tables 2, 3).

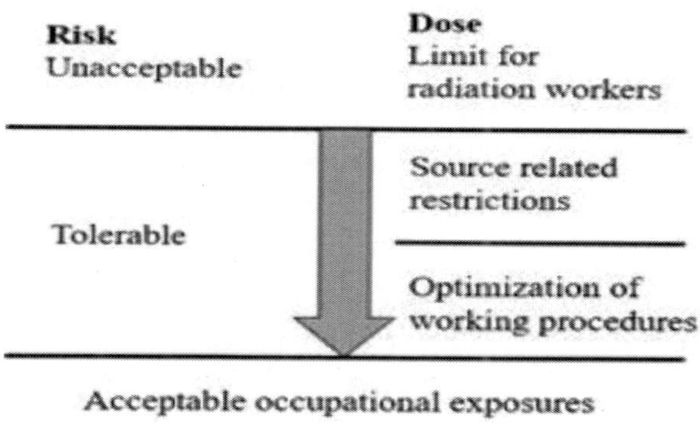

Figure 5. Optimisation in radiation protection.

Individuals for whom doses have been optimised using ALARA principles should not be exposed to radiation above the established dose limits. The nationally and internationally accepted maximum limit value for members of the public is 1 mSv per year. The international limit for radiation workers is 100 mSv in total over five years (not to exceed 50 mSv per year). In recent years, these dose limits have been reduced, and it has been recommended that an annual value of 20 mSv should be applied for workers. In emergency cases, one-time exposure is 50 mSv provided that over 5 consecutive years, the average dose does not transgress 20 mSv dose.

5.3.1. Dosage Limits

They are grouped as primary, secondary, derived, Permissible, and Operating Limits in addition to dose equivalent indices. The most commonly used limits for those working with ionising radiation and for the public are *Primary* and *Secondary Limits*.

5.3.1.1. Primary Limits

These are the annual dose limits that radiation officials or members of the public can receive. The restriction is that they do not allow routine practices to be carried out above these limits.

5.3.1.2. Secondary Limits

These are the limits for the possibility of internal irradiation, and they are determined in tables for each radionuclide. The definition of Annual Limits on Intake (ALI) is important in these limits. ALI is defined as the annual limit values of radioactive substances that can be taken into the body in Becquerel (Bq).

For young workers the following points provide importance:

- People under the age of 16 cannot be employed in jobs that will expose them to occupational exposure.
- People under the age of 18 cannot be employed in controlled areas unless supervised and for educational purposes.

On the other hand, for pregnant employees, the following rules must be considered:

- Dose limits are not different for men or women under normal working conditions.
- Working conditions should be adjusted to ensure that the embryo or fetus for protected at a level not exceeding the level allowed for the public (1mSv).

In external irradiation after diagnosis of pregnancy, the dose limit for the abdominal surface (until the end of pregnancy) is 2mSv, and for internal irradiations, it is 1/20 as recommended by ALI.

For any justified application, radiation protection must be optimized. Irradiation of individuals should be at the lowest possible dose and below international regulatory limits (Zeyrek, 2013). However, the dose received by individuals should not exceed the equivalent dose in the relevant organ or tissue caused by all permitted irradiation and the effective dose values, which are received in natural radiation and medical irradiation as not taken into account in applying dose limits (IAEA BSS 115, 1996; Sowby, 1981b). ICRP continues its work collaboration with other international radiation protection organizations and in its latest reports, ICRP has defined the expression "*Maximum Permissible Doses*" to reduce human radiation exposure and these values are updated within the framework of research (ICRP 9, 2006; Çimen, 2017).

6. Application of the Principles of Radiation Protection

The principles of radiation safety are simple and easy to learn and remember, but the main thing is that the application of these principles is based on some knowledge or guesses about the radiological conditions.

Table 5. Expanded radiation protection principles

Principle	What to Do as an Individual	Technical Requirements
Time	Hurry up without making a fuss. Finish your work as soon as possible	Exposure to ionising radiation/short stay near radiation
Distance	Stay away from radiation as much as possible.	The intensity of radiation decreases inversely proportional to the square of the distance. Maximize distance.
Dispersion	Scatter and dilute radiation.	Minimize concentration, maximize dilution.
Source Reduction	Use as little ionising radiation as possible	Minimizing the production and use of radiation and radioactive substances.
Armouring	Devices that emit ionising rays should be kept in armoured rooms within the framework of protocols.	Maximize absorption (shield)
Personal Armouring	personal protection (lead vest, thyroid protection, gloves, etc.) with you and use it.	Minimize (control and limit) emissions
Decontamination (internal and external irradiation)	Let the radiation environment be outside	Minimizing the entry of radiation and radioactive substances into the body
Mitigation	Take off contaminated clothing. Try to remove ionising radiation with chemicals in internal irradiation.	Maximize removal or inhibition of materials from the body (after ingestion or skin contamination)
Optimal technology	Work to limit the damage.	Optimize exposure over time and between individuals, scavenge free radicals, and promote repair.
Other (Limiting radiation exposure)	Choose the best and safest technology. Investigate and escalate risks	Optimize risk-benefit-cost ratio Minimize exposure to other agents

The basic principles of radiation safety are discussed in detail in Daniel Strom's article (Strom D., 1996). In his article, Daniel Strom mentions that "time, distance, and shielding," which form the basis of the principles, are enough for protection from external radiation, but are insufficient against many types of radiation such as internal irradiation and radon gas present in closed areas.

An adapted summary of these principles can be seen in Table 5. The dispersion principle in the table is valid for radioactive substances in air, water, or soil.

Technically, if the concentration is minimised and dilution is maximized, the dose that can be taken will also decrease. The article mentions that the application areas of this principle are radon formation in mines/homes and medical situations, where patients may inhale radioactive gas. Additionally, Daniel Strom argues that the principle of distribution as a disadvantage is the opposite of the resource barrier principle.

6.1. Radon Gas

The largest contribution to human radiation exposure from natural sources comes from radon gas (Rn), which is released from the soil and accumulates in dwellings (Figure 6). It occurs naturally originating from a series of uranium (U) decay and radium (226Ra).

As a noble gas, radon is imperceptible to human senses and freely diffuses into the atmosphere. Uranium is a natural part of the earth's crust. For this reason, radium and radon are naturally present in the environment. Radon decays with a half-life of 3.8 days.

These short half-life isotopes are chemically active solid particles that cling to suspended particles in the air and cannot be detected by human senses. They settle in the lungs through inhalation and cause the lungs to be irradiated and dosed with the alpha particles they emit (UNSCEAR, 1988).

It is known that radon is the second most common cause of lung cancer after smoking. Epidemiological studies prove a correlation between indoor radon exposure and lung cancer, even though radon levels are relatively low in residences (NRC,1999). The World Health Organization first drew attention to the health effects of radon exposure in homes with the European working group on indoor air quality established in 1979.

In many countries of the world, studies to determine radon concentration levels throughout the country have been published (WHO, 2009). Exposure to radon is limited by the International Commission on Radiological Protection (ICRP) and an action level has been recommended for the annual mean radon concentration in buildings (ICRP, 1993).

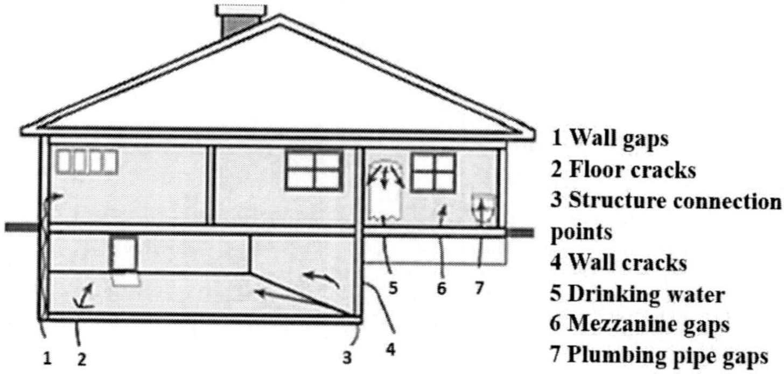

Figure 6. Ways radon enters the home environment.

7. Radiation Protection in the Nuclear Industry

Since the development of civilian nuclear power in the 1950s, relatively few accidents have occurred with off-site consequences. Such incidents are rare due to efforts to implement the necessary preventive, protective, and mitigating safety measures for all types of nuclear facilities. These measures cover structures, systems, and components (SSC), as well as the management of the facility. The potential for accidents that may have high consequences causes the high financial costs observed in nuclear safety to increase, even if they have very low probabilities (Ashley et al., 2017). First responders and other rescuers to a radiological attack must be adequately trained and have appropriate equipment (Geiger-Müller - GM) personal protectors, appropriate dosimeters, etc. to detect radiation and radioactive contamination. Additionally, radiological protection experts must be available to advise in all situations. It is safe to assume that radiological, chemical, and/or biological agents are all involved in the attack until proven otherwise. ICRP defines exposures as normal times that allow for planned control (planned exposures), emergencies such as accidents or nuclear terrorism (emergency exposures), and the recovery and reconstruction period after an accident (current exposures), and each of these is established according to protection standards for one (ICRP 103).

Planned exposure situations, Preventive measures used in normal times should aim to prevent any exposure and reduce the risk of developing cancer in the future as much as possible. In no case should the dose limit of 1 mSv determined for annual public exposure be exceeded. This limit indicates that places where radiation or radioactive substances are processed must be operated in accordance with the rules to prevent the annual exposure dose level of the public from being exceeded (J. Valentin, 2007).

On the other hand, in emergency exposure situations such as nuclear accidents, physical disorders that cannot be seen in normal times may develop, so measures should be taken to prevent serious physical disorders rather than measures to reduce the risk of developing cancer in the future. For this reason, instead of applying the standard dose limits of 1 mSv/year for the public, it is more appropriate to carry out studies to reduce leakage doses by increasing these limits to the reference level of 20 to 100 mSv/year. Previous studies recommend that post-emergency dose values should be reduced during the recovery and reconstruction period to a reference level in the range of 1 to 20 mSv/year, which is lower than the emergency reference level but higher than the applicable dose limits. In epidemiological studies on exposure to low doses and low dose rates, the results of animal tests and experiments using human cells for chromosomal induction indicate that cancer is triggered at doses below 100 mSv, which is the defined threshold dose (Lubin et al., 2017; Veiga et al., 2016).

The main purpose of radiological protection after an attack is to prevent the emergence of acute radiation effects (deterministic effects) due to radiation exposure and to limit the possibility of late effects (stochastic effects) such as cancers and some hereditary diseases. In addition, it is necessary to minimize environmental contamination from radioactive residues and the resulting disruption of daily life. Reports mentioned that measures to prevent exposures are a much more effective preventive measure (ICRP 113, NCRP 116). Responders involved in rescue, recovery, and restoration efforts must be familiar with and apply the usual international standards for occupational radiological protection based on ICRP recommendations (ICRP-96, 2005).

Radiological protection is the most important safety issue of the nuclear industry because uranium isotopes and nuclear fission products emit radiation. Different radiation protection problems are encountered at each stage of nuclear fuel use. For example, workers in uranium mining are exposed to dust from uranium and its products. This dust is potentially dangerous when inhaled into the lungs. For this reason, the mine must have an appropriate ventilation system and a high level of protection for workers' breathing (ICRP-96, 2005). For those working in nuclear power stations, exposure is due to gamma-emitting radionuclides (such as Cobalt-60).

During normal operating hours, these systems are in armoured areas and employees are outside the hazardous radiation area. Workers may be exposed to radiation during the maintenance of systems in the power plant (such as pumps and reactor cooling water systems). Protection of employees from radiation during maintenance is ensured using appropriate armour systems and appropriate selection of job responsibilities (TAEK, 2012; Hedemann-Jensen & McEwan, 2011).

8. Radiation Protection (Monitoring)

Monitoring is to determine the presence and degree of ionising radiation and radioactive contamination. Radiation monitoring is an integral part of the radiation protection programme. There are two types of monitoring: area and personnel (Zeyrek, 2013).

8.1. Personnel Monitoring

Total body intake by individuals' routine dose is measured as indirect and direct monitoring devices.

8.1.1. Indirect Monitoring

Dose readings are devices that require an additional device and measure the total radiation dose received over a certain period (Film Dosimeters, TLD Dosimeters, OSL Dosimeters).

8.1.2. Direct Monitoring

These are direct reading devices (Pocket (Pen) Dosimeters, Electronic Dosimeters) that make it possible to directly read the radiation doses received.

8.2. Objectives of Personnel Monitoring Service

To keep the personal radiation doses exposed to the personnel below the maximum permissible level, measuring and keeping records of the doses received, giving assurance to the radiation worker that his health is protected in terms of radiation, and providing the opportunity for legal determination of overdose disputes between the organisation and the personnel.

8.2.2. Area Monitoring

The purpose of area monitoring is to prevent unnecessary exposure of individuals to radiation and radioactive substances (safety) and to prevent radioactive substances from being accessed by unauthorized persons (for safety reasons). Especially, areas, where the annual dose to be exposed is likely to exceed 1 mSv, are considered as radiation areas, which are classified according to radiation levels (Zeyrek, 2013).

8.2.2.1. Controlled Areas

These are areas where people who work with radiation as part of their duties may be exposed to a radiation dose of more than 3/10 of the average annual dose limits for five consecutive years. Entrances and exits of radiation officers in these areas are subject to special control. Works are subject to special rules regarding radiation protection.

8.2.2.2. Supervised Areas

For radiation officers, 1/20 of the annual dose limits are likely to be exceeded, and 3/10 are areas, where it is not expected to be exceeded. In these areas, personal dose measurement is not required but monitoring of environmental radiation is necessary.

8.3. Visitors

Visitors can neither enter controlled areas nor can they enter supervised areas without permission from the radiation protection officer.

8.4. Controlled Areas and Their Entrances

Basic radiation signs indicate that it is a radiation area, signs bearing the necessary information, symbols, and colours show the magnitude and characteristics of the danger of radiation exposure in an understandable way. Warning signs should indicate that protective clothing and tools should be used, and necessary radiation measurement devices and personal dosimeters should be used. Haematological and skin and eye examinations of employees are performed at least once a year.

Conclusion

Since people cannot fully control their radiation exposure, they can keep it under control with dose limits and risk assessments. The aim here is to ensure that no one, including future generations, is exposed to the risk of unwanted ionising radiation. The applied system aims to protect radiation officials, the public, and the environment from the harmful effects of ionising radiation. To ensure radiation protection and safety, it is necessary to establish and maintain a relevant system at national and international levels. The method of protection from radiation requires knowing the quantities and units related to the radiation dose well, making the necessary measurements, and evaluating the measurement results.

References

Ashley, S. F., Vaughan, G. J., Nuttall, W. J., & Thomas, P. J. (2017). Considerations in relation to off-site emergency procedures and response for nuclear accidents. *Process Safety and Environmental Protection*, 112, 77–95. https://doi.org/10.1016/j.psep.2017.08.031.

Bahattin Ç., Erdoğan M., R. O. (2017). Ionizing Radiation and Protection Methods. *Üniversitesi, Selçuk Fakültesi, Fen Bölümü, Fizik*, *43*(2), 139–147. www.iaea.org.

Charles B. Meinhold. (1993). *NCRP REPORT No. 116:* Limitation of Exposure to Ionizing Radiation. *NCRP 1993*.

Do, K. H. (2016). General principles of radiation protection in fields of diagnostic medical exposure. *Journal of Korean Medical Science*, *31*, S6–S9. https://doi.org/10.3346/
jkms.2016.31.S1.S6.

Donald A. Cool. (2015). *Application of the Commission's Recommendations Program of Work. October*.

Farmelo, G. (1995). The Discovery of X-rays. *Scientific American*, November.

Gökharman, D. F., Aydın, S., & Koşar, P. N. (2016). Need To Know About Radiation Safety Vocationally. *Sdü Sağl Bilimleri Dergisi*, *7*(2), 35–35. https://doi.org/ 10.22312/sdusbed.261237.

Griffey, R. T., & Sodickson, A. (2009). Cumulative radiation exposure and cancer risk estimates in emergency department patients undergoing repeat or multiple CT. *American Journal of Roentgenology*, *192*(4), 887–892. https://doi.org/10.2214/ AJR.08.1351.

Hedemann-Jensen, P., & McEwan, A. C. (2011). Is ICRP guidance on the use of reference levels consistent? *Journal of Radiological Protection*, *31*(4), 431–444. https://doi.org/
10.1088/0952-4746/31/4/003.

Holmberg, O., Malone, J., Rehani, M., McLean, D., & Czarwinski, R. (2010). Current issues and actions in radiation protection of patients. *European Journal of Radiology*, *76*(1), 15–19. https://doi.org/10.1016/j.ejrad. 2010. 06.033.

IAEA BSS 115. (1996). Categories in the IAEA Safety Series. *IAEA*, 48. http://www.ilo.org/wcmsp5/groups/public/ed_protect/protrav/safework/documents/ publication/wcms_152685.pdf.

ICRP 103. (2007). *The 2007 Recommendations of the International Commission on Radiological Protection* (J. Valentin (ed.)). ICRP By Elsevier. http://www.elsevier. com/wps/find/bookdescription.cws_home/713998/description#description.

ICRP 26. (1979). The 1978 Stockholm meeting of the International Commission on Radiological Protection. *Acta Oncologica*, *18*(1), 77–80. https://doi.org/ 10.3109/02841867909128193.

ICRP 9. (2006). The 2006 recommendations of the International Commission on radiological protection. In *Proceedings of the 31st World Nuclear Association Annual Symposium, 2006* (Vol. 2006). Pergamon Press- Icrp.

ICRP-96. (2005). Protecting People against Radiation Exposure in the Event of a Radiological Attack. *ICRP by Elsevier.*

International Atomic Energy Agency (IAEA). (2023). *IAEA* Human Health Series Publications*: Handbook of Basic Quality Control Tests for Diagnostic Radiology*. *47*. http://www.iaea.org/Publications/index.html.

Kachelrieß, M., & Rehani, M. M. (2020). Is it possible to kill the radiation risk issue in computed tomography? *Physica Medica*, *71*, 176–177. https://doi.org/10.1016/ j.ejmp.2020.02.017.

Kase, K. R. (2004). Radiation protection principles of NCRP. *Health Physics*, *87*(3), 251–257. https://doi.org/10.1097/00004032-200409000-00005.

Khan, F. M. (2003). Physics of Radiation Therapy Third Edition. *The Journal of the American Medical Association*, 1138.

Lubin, J. H., Adams, M. J., Shore, R., Holmberg, E., Schneider, A. B., Hawkins, M. M., Robison, L. L., Inskip, P. D., Lundell, M., Johansson, R., Kleinerman, R. A., De Vathaire, F., Damber, L., Sadetzki, S., Tucker, M., Sakata, R., & Veiga, L. H. S. (2017). Thyroid cancer following childhood low-dose radiation exposure: A pooled analysis of nine cohorts. *Journal of Clinical Endocrinology and Metabolism*, *102*(7), 2575–2583. https://doi.org/10.1210/jc.2016-3529.

Mettler, F. A., Mahesh, M., Bhargavan-Chatfield, M., Chambers, C. E., Elee, J. G., Frush, D. P., Miller, D. L., Royal, H. D., Milano, M. T., Spelic, D. C., Ansari, A. J., Bolch, W. E., Guebert, G. M., Sherrier, R. H., Smith, J. M., & Vetter, R. J. (2020). Patient exposure from radiologic and nuclear medicine procedures in the United States: Procedure volume and effective dose for the period 2006–2016. *Radiology*, *295*(2), 418–427. https://doi.org/ 10.1148/radiol.2020192256.

NCRP 91. (1987). Recommendations On Limits For Exposure To Ionizing Radiation. (Vol. 28). *NCRP.*

Ozturk, N. (2020). Measurement of Ionizing Radiation and Radiation Dose Units. In A. F. Becerir H. B. (Ed.), *Radiotherapy physics* (Vol. 7, pp. 83–111). İstanbul Nobel Akademik Yayıncılık Eğ Danışmanlık Tic Ltd. Şti.

Radiation Protection N° 180. (n.d.). European Comission Radiation Protection N° 180 *Medical Radiation Exposure of the European Population Part 1/2*. https://ec.europa.eu/energy/sites/ener/files/documents/RP180.pdf.

Rehani, M. M., Szczykutowicz, T. P., & Zaidi, H. (2020). CT is still not a low-dose imaging modality. *Medical Physics*, *47*(2), 293–296. https://doi.org/10.1002/mp.14000.

Salomaa, S., Jourdain, J. R., Kreuzer, M., Jung, T., & Repussard, J. (2017). Multidisciplinary European low dose initiative: An update of the Melodi program. *International Journal of Radiation Biology*, *93*(10), 1035–1039. https://doi.org/10.1080/09553002.2017.1281463.

Schauer, D. A., & Linton, O. W. (2009). NCRP Report No. 160, ionizing radiation exposure of the population of the United States, medical exposure - Are we doing less with more, and is there a role for health physicists? *Health Physics*, *97*(1), 1–5. https://doi.org/10.1097/01.HP. 0000356672.44380.b7.

Sebastian Pflugbeil, H. P. C. (2011). *Chernobyl's Effects on Human Health 25 Years After the Reactor Disaster.*

Sowby, F. D. (1981a). Annals of the ICRP-60. *Annals of the ICRP*, *6*(1), 1. https://doi.org/10.1016/0146-6453(81)90127-5.

Sowby, F. D. (1981b). Annals of the ICRP. *Annals of the ICRP*, *6*(1), 1. https://doi.org/10.1016/0146-6453(81)90127-5.

STROM. D. J. (1996). Strom1996_PNWD-SA-4290_Ten_Principles_and_Ten_Commandments_of_Radiation_Protection_OCR.pdf. *Health Physics*, *7*(3), 388–393.

TAEK. (2012). Radyasyondan koruyucu donanimlara ilişkin kılavuz. *TAEK* 1–9.

UNSCEAR. (1988). Source, effects, and risks of ionising radiation - exposures from natural sources of radiation. *Health Physics*, 647. https://www.unscear.org/docs/publications/
1988/UNSCEAR_1988_Report.pdf.

UNSCEAR. (2000). Sources and Effects of Ionizing Radiation Volume I: source. In *United Nations Scientific Committe on the Effects of Atomic Radiation: Vol. I.*

UNSCEAR. (2015). Sources, effects, and risks of ionizing radiation, 2012 Report to the General Assembly, Scientific Annexes A and B. In *United Nations Publication*. https://www.unscear.org/docs/publications/2012/ UNSCEAR_ 2012_Report.pdf.

UNSCEAR. (2016). Unscear2015 Developments Since The 2013 Unscear Report On The Levels And Effects Of Radiation Exposure Due To The Nuclear Accident Following The Great East-Japan Earthquake And Tsunami. *United Nations Publication*. www.unscear.org.

Valentin, J. (2007). Radiation protection in medicine. ICRP Publication 105. *Annals of the ICRP 105*, *37*(6), 1–63. https://doi.org/10.1016/j.icrp. 2008.07.001.

Vano, E., Miller, D. L., Martin, C. J., Rehani, M. M., Kang, K., Rosenstein, M., Ortiz, P., Mattsson, S., Padovani, R., & Rogers, A. (2016). *Annals of the ICRP-60. 45*(2), 1–73.

Veiga, L. H. S., Holmberg, E., Anderson, H., Pottern, L., Sadetzki, S., Adams, M. J., Sakata, R., Schneider, A. B., Inskip, P., Bhatti, P., Johansson, R., Neta, G., Shore, R., de Vathaire, F., Damber, L., Kleinerman, R., Hawkins, M. M., Tucker, M., Lundell, M., & Lubin, J. H. (2016). Thyroid Cancer after Childhood Exposure to External Radiation: An Updated Pooled Analysis of 12 Studies. *Radiation Research*, *185*(5), 473. https://doi.org/10.1667 /RR14213.1.

Wojcik, A., & Harms-Ringdahl, M. (2019). Radiation protection biology then and now. *International Journal of Radiation Biology*, *95*(7), 841–850. https://doi.org/10.1080/ 09553002.2019.1589027.

Zeyrek, C. T. (2013). General Concepts of Safety and Protection for Ionizing Radiation Applications. *Süleyman Demirel Ü. Fen Bilm Enst. D*, *17*(3), 1-9.

Chapter 3

Biological Effects of Radiation

Naciye Işık*

Radiation Oncology, Health Sciences University Kartal Dr. Lütfi Kırdar City Hospital
Istanbul, Türkiye

Abstract

Several factors influence the biological effects of radiation, including the type of radiation, its energy, the received dose, the duration of exposure, and the sensitivity of the tissues or organs exposed. The impact of radiation on biological tissues is generally classified into stochastic and deterministic effects. At the cellular level, radiation can cause direct and indirect harm, potentially leading to membrane disruption, cell death, altered enzyme activity, metabolic issues, mitochondrial dysfunction, accelerated aging, genetic instability, and mutagenesis. Acute tissue damage from radiation may present as inflammation, edema, vascular injury, and necrosis, whereas chronic exposure might result in fibrosis, organ failure, and a heightened cancer risk. Radiation-induced tissue damage can initiate inflammatory responses marked by the secretion of pro-inflammatory cytokines, chemokines, and other signaling molecules. Additionally, radiation exposure can cause oxidative stress and epigenetic changes in healthy tissues. Ionising radiation damages cellular DNA, leading to mutations that can cause uncontrolled cell growth and tumour formation, attributed to its oncogenic potential. Comprehending these effects is vital for evaluating radiation exposure risks and formulating protection and mitigation strategies.

This chapter discusses the damage caused by radiation to living cells and tissues.

---

* Corresponding Author's Email: nacisik@hotmail.com.

In: Triage in Radiation Accidents
Editors: Nural Öztürk and Sibel Karaca
ISBN: 979-8-89530-415-0

Keywords: ionising radiation, radiation effects, tissue reactions, deterministic effect, stochastic effects

1. Introduction

Various factors lead to biological effects, such as the type of radiation, its energy, the dose received, the duration of exposure, and the sensitivity of exposed tissues or organs. Understanding these effects is crucial for assessing the risks associated with radiation exposure and developing strategies for radiation protection and mitigation.

Radiation's impact on biological systems is a complex interplay of factors, with ionisation density and linear energy transfer (LET) as crucial determinants (Nikitaki et al., 2016; Hada et al., 2011). LET, which measures the energy absorbed over a certain distance, varies with different types of radiation, influencing the extent of biological damage. Low-LET radiation, like X-rays, distributes energy evenly, causing widespread but generally less severe effects (Crompton et al., 2018). High-LET radiation, such as alpha particles, delivers energy rapidly, producing more concentrated and severe damage (Donya et al., 2015; Mehta, 2005). Understanding these differences is vital for assessing risks and implementing safety measures in environments where radiation exposure is possible.

The biological effects of radiation encompass a wide range of responses in living organisms exposed to ionising radiation. Radiation effects on biological tissues can be categorised into stochastic and deterministic effects (Vetter, 2008). Deterministic effects also referred to as non-stochastic effects, exhibit a dose-response relationship where the effect's severity increases with the dose's magnitude. Stochastic effects are probabilistic and do not exhibit a threshold dose. Understanding these effects is crucial in radiology, nuclear medicine, and radiation oncology.

2. Deterministic Effects

Predictably, these effects can be anticipated based on the dose received, leading to immediate or short-term health consequences post-exposure. Deterministic effects are typically associated with relatively high doses of radiation delivered over a short period, such as those encountered in radiation therapy for cancer treatment or nuclear accidents. The severity of these effects depends on factors such as the type of radiation, the dose rate, the duration of exposure, and the sensitivity of the exposed tissues and organs. The predictable nature of deterministic effects and their dose-dependent severity necessitates stringent dose limits and protective measures to ensure that radiation exposures remain below threshold levels (Ban et al., 2022). Deterministic effects are crucial in medical radiology, radiation therapy, and occupational exposure settings to prevent acute and chronic health detriments.

2.1. Cellular Damage

Radiation cellular damage occurs when ionising radiation interacts with the cells of living organisms, disrupting their normal biological processes and structures. Radiation creates a biological effect by damaging a Deoxyribonucleic acid (DNA) molecule that contains genetic information for proliferation in the cell or a cellular component necessary for the cell to survive. (Einor et al., 2016; Yahyapour et al., 2018). Radiation's most evident cellular impact is its ability to cause direct or indirect DNA breaks, leading to apoptosis and subsequent cell death.

Direct effects result from direct ionisation and excitation of DNA molecules (Sies, 2015). The indirect effect is the free radicals formed from water and oxygen in the tissue that cause the effects of DNA damage. The initial step in forming a free radical during water radiolysis is the emission of an electron when ionising radiation interacts with a water molecule (J Ramos-Méndez et al., 2021). This process produces reactive oxygen species (ROS) such as the superoxide anion, hydroxyl radical, hydrated electron, and hydrogen peroxide. Additionally, the release of nitric oxide and its metabolites, including peroxynitrite and nitrogen dioxide, plays a role in ionising radiation-induced genomic damage. These reactive molecules can oxidize and damage DNA, proteins, lipids, and other cellular components, leading to oxidative stress and cell dysfunction (Einor et al., 2016; Wardman, 2009; Sies, 2015).

Ionising radiation can directly or indirectly interact with the DNA molecules within cells, causing various types of damage, including single-strand breaks, double-strand breaks, and cross-linking of DNA strands (Prise et al., 2001). Furthermore, the formation of DNA adducts can hinder DNA repair mechanisms. Detection of radiation-induced DNA damage involves the kinases ataxia-telangiectasia mutated (ATM) and ataxia-telangiectasia

Rad3-related protein (ATR) (Culligan et al., 2006; Marechal & Zou, 2013). The cell cycle stops, and repair mechanisms are activated. Non-homologous end joining (NHEJ) misrepair is possible but is active throughout the cell cycle; Homologous recombination (HR), on the other hand, is error-free but requires a repair template, so it can only be used in the late S and G2 phases of mitosis. Base Excision Repair (BER) and Nucleotide Excision Repair (NER) processes fix base modifications and bulky lesions, respectively (Cooper & Hausman, 2000). Errors during these processes can result in point mutations or small insertions/deletions. If damage repair is not possible, cell death will occur at varying times after radiation exposure through one of the mechanisms of mitotic catastrophe, apoptosis, senescence, necrosis, or autophagy (Sia et al., 2020).

Cells exposed to radiation may undergo cell cycle arrest, halting their progression through the cell cycle to allow time for DNA repair or triggering apoptosis if the damage is too severe to repair. Failure to repair DNA damage or aberrant activation of apoptosis pathways can lead to genomic instability. These DNA lesions can interfere with DNA replication and transcription, leading to mutations and chromosomal aberrations. Radiation-induced DNA damage and repair processes can result in genomic instability, characterised by an increased propensity to accumulate additional mutations and chromosomal rearrangements over time. Genomic instability is a hallmark of radiation-induced carcinogenesis (Little, 2009; Morgan, 2011; Ozasa et al., 2012).

Research continues to understand the changes in the microenvironment of cellular components exposed to radiation and their role in tissue and organ dysfunction following irradiation. Non-nuclear subcellular structures are known to be important but have yet to be adequately explained (Afshari et al., 2024). Cells damaged by ionising radiation can induce radiation bystander effects in non-irradiated cells, leading to various changes such as alterations in gene expression, protein synthesis, chromosomal aberrations, formation of micronuclei, secretion of exosomes and microRNAs, as well as cell death, proliferation, or transformation (Olivares et al., 2021). Reactive oxygen species are considered as the initiators of these effects, while nitric oxide (NO), transforming growth factor beta (TGF-β), and other inflammatory cytokines act as effectors (Klammer et al., 2015). Furthermore, the inflammatory response perpetuates a cycle of ROS, cytokines, chemokines, and growth factors, accompanied by inflammatory infiltrates, creating a feedback loop where oxidative stress and inflammation perpetuate each other (Afshari et al., 2024; Klammer et al., 2015; Olivares et al., 2021).

Cell death by radiation depends on many factors, such as cell radiosensitivity, radiation dose and type, oxygen availability, the genetic antitumour activity of the cell, DNA repair capacity, location in the cell cycle during radiation exposure, and microenvironment (Sia et al., 2020). While the direct effect is in the foreground in high LET radiation, the indirect effect is in the foreground in low LET radiation. (Nikitaki et al., 2016; Méndez et al., 2021). Not all cells are affected by radiation to the same extent. Radiosensitivity depends on proliferation and repair potential. Cells that divide more rapidly have less repair potential and are more susceptible to damage. The extent and severity of cellular damage caused by radiation depend on factors such as the type and energy of radiation, the dose received, the dose rate, the duration of exposure, and the radiosensitivity of the cells or tissues involved (Hall & Giaccia, 2019). Protective mechanisms such as DNA repair pathways, antioxidant defenses, and cell cycle checkpoints help mitigate radiation-induced damage and maintain cellular homeostasis. The effects of radiation include repairing sublethal damage and replacing lethally injured cells through repopulation over extended periods. Other intracellular repair mechanisms, such as 'potentially lethal damage' and 'slow repair,' may also enhance survival (Hall & Giaccia, 2019). Additionally, cell replacement can occur through the migration of unirradiated cells from unaffected areas (Olivares et al., 2021). While higher total doses are typically more tolerable after prolonged or chronic exposure than acute exposure, the injury mechanisms may vary depending on the dose and pattern of dose delivery. However, excessive or prolonged exposure to radiation can overwhelm protective mechanisms, leading to adverse biological effects (Hendry, 2012).

2.2. Tissue Injury

Radiation tissue damage is a complex and multifaceted process that involves interactions between ionising radiation and biological tissues at the cellular, molecular, and tissue levels. Different tissues and organs have varying sensitivities to radiation (Hall & Giaccia, 2019). For example, rapidly dividing cells, like those in the bone marrow and the gastrointestinal tract lining, are more susceptible to radiation damage. The structural composition of tissues and organs is increasingly recognised as a crucial factor in their response to radiation. Organs with paired or parallel-arranged functional subunits, such as kidneys and lungs, can endure the inactivation of numerous subunits without manifesting clinical damage due to their significant reserve capacity and the compensatory abilities of the remaining units. Conversely, organs structured serially, like the spinal cord, possess minimal or no reserve capacity, making their tolerance dose less reliant on the irradiated volume. In such organs, functional impairment post-threshold dose typically presents as a binary outcome rather than a gradation of severity with increased dose (Hall & Giaccia, 2019).

The timing of biological effects following radiation exposure can vary depending on several factors, including the type of radiation, the dose received, the dose rate, and the sensitivity of the tissues or organs involved. Early effects of radiation exposure may occur within weeks after exposure, particularly in tissues with high cell turnover rates. Early tissue reactions may be inflammatory due to changes in cell permeability and the release of inflammatory mediators (Garau et al., 2011). These reactions often stem from cell loss and various acute radiation syndromes, though non-cytotoxic effects on tissues also play a role.

Late tissue reactions are categorised as 'direct' when they stem from damage to the targeted tissue, or as 'consequential' when they arise from severe initial reactions. The late effects of radiation exposure usually appear months to years or even decades post-exposure and can lead to chronic conditions like cancer, fibrosis, and heart disease. Fibrosis, marked by scar tissue formation in irradiated areas, can arise months to years after exposure, potentially affecting tissue function. These late effects of radiation often progress and become more common as time passes. Research indicates that late tissue damage follows an exponential pattern, measurable by the yearly percentage of patients at risk for a specific effect (Jung et al., 2001). Emphasises the necessity for long-term follow-ups to evaluate the lifelong cumulative impact of such injuries on workers and the public.

Acute tissue injury may manifest as inflammation, edema, vascular damage, and cell death, while chronic exposure can lead to fibrosis, organ dysfunction, and an increased risk of cancer. Radiation-induced tissue damage can trigger inflammatory responses in irradiated tissues by releasing pro-inflammatory cytokines, chemokines, and other signaling molecules (Wardman, 2009). Radiation exposure can damage blood vessels and impair blood flow to affected tissues, leading to ischemia and tissue necrosis. Vascular damage can contribute to the development of radiation-induced fibrosis and compromise tissue healing and repair processes. Chronic radiation exposure can lead to the accumulation of extracellular matrix proteins and the formation of fibrosis in irradiated tissues. Radiation-induced fibrosis occurs with the activation of myofibroblasts through intracellular and extracellular signals. It occurs through oxidative stress response, metabolic reprogramming, proteasome pathway activation by immune cells, and abnormal cytokine release (Yu et al., 2023). Radiation-induced tissue fibrosis can occur in various organs, including the lungs, skin, liver, and kidneys.

2.3. Effect of Radiation on Systems

2.3.1. Respiratory System

Exposure of lung tissue to ionising radiation damages alveolar epithelial cells and vascular endothelial cells, producing large quantities of cytokines that cause acute radiation pneumonia. This condition prompts the release of additional inflammatory cells due to cytokines such as TGF-β, IFN-γ, ET-1, IL-4, and IL-13, as well as chemokines and growth factors (Huang et al., 2017).

Exposure to radioactive radiation can have significant adverse effects on the respiratory system. Potential outcomes include Acute Radiation Syndrome, radiation pneumonitis, a heightened risk of developing lung cancer, susceptibility to respiratory infections, and persistent respiratory complications. Chronic damage from this process can eventually result in pulmonary fibrosis (Huang et al., 2017; Shin et al., 2020).

2.3.2. Cardiovascular System

An increasing amount of epidemiological evidence indicates higher risks of cardiovascular diseases, notably ischemic heart disease and stroke, associated with radiation. (Little et al., 2012; Peters et al., 2023). High levels of radiation exceeding 10 Gy can lead to acute microvascular disease, which damages small blood vessels, as well as chronic macrovascular disease, affecting the coronary arteries. The primary biological mechanism for these diseases is inflammation, resulting in pro-inflammatory signaling pathways and atherosclerotic changes. These pathological alterations are similar to those observed in ischemic heart disease caused by a range of other factors (Tapio, 2016). The impact on cardiovascular health needs to be more clearly defined at lower doses, specifically moderate to low doses. Potential mechanisms for damage may involve harm to endothelial cells and the initiation of pro-inflammatory responses. Exposure of healthy endothelial cells to low-dose radiation can lead to the production of intracellular reactive oxygen species and subsequent oxidative stress. This process has the potential to hasten vascular inflammation and contribute to the development of atherosclerotic plaques (Tapio, 2016; Little et al., 2012; Tapio et al., 2021).

2.3.3. Central Nervous System

Reports have indicated associations between exposure to radiation and neurological effects on the brain (Lopes et al., 2022; Tohidinezhad et al., 2022; Jayan et al., 2022). Vascular lesions may result in cerebral vascular syndrome, manifesting hours after exposure. Significant cerebral edema due to vascular damage can lead to headaches, nausea, seizures, herniation, and potentially, death. Acute radiation injury is primarily caused by demyelination and edema. (Acosta and Warrington, 2020). Acute encephalopathy, marked by temporary fluid buildup in brain cells resulting in cerebral edema, may arise within two weeks following radiation. Early-delayed radiation encephalopathy, which can start from two weeks to six months post-radiation therapy, is likely due to temporary demyelination from oligodendroglial damage and brain barrier disruption, with symptoms akin to acute encephalopathy (Acosta & Warrington, 2020). Long-term effects encompass changes in white matter and neuronal loss. Late complications, such as radionecrosis and late-delayed radiation encephalopathy, generally appear 6 to 12 months after radiation, with radionecrosis being one of the gravest side effects of radiotherapy (Tohidinezhad et al., 2022). Late-delayed radiation encephalopathy may cause cognitive deficits ranging from mild memory problems to severe dementia (Srivastava et al., 2023). A higher dose-risk relationship has been observed for the incidence and mortality of cerebrovascular diseases (Lopes et al., 2022).

2.3.4. Gastrointestinal Tract

Radiation can impact the intestines, potentially causing damage to various intestinal cells. This damage is primarily attributed to the apoptosis and necrosis of epithelial and endothelial cells (Shadad, 2013). The mucosal dysfunction of acute radiation enteritis is commonly associated with the impact of ionising radiation on the cell cycle of intestinal cryptepithelial stem cells, resulting in epithelial loss (Lu et al., 2023). Nonetheless, recent discoveries suggest that preceding events might also play a role. The severity of these early events could affect the development and severity of chronic enteritis. While the precise mechanism remains unclear, it may involve a radiation-induced impairment of the host's defense against luminal pathogens or antigens (MacNaughton, 2000). The intestine, which regenerates rapidly and is highly sensitive to radiation, can experience chronic dysfunction following abdominal/pelvic radiation, leading to symptoms such as vomiting, weight loss, anorexia, dehydration, diarrhea, and infections. In severe cases, this may lead to intestinal necrosis, perforation, and even death (Garau et al., 2011).

2.3.5. Hematopoietic System

Irradiation can lead to a hematopoietic syndrome, a consequence of stem cell damage in the bone marrow and lymphatic tissues. Even lower doses of radiation can cause a minor reduction in blood cell count (Henry & Arcangeli, 2020). Significant alterations in blood composition occur, including pronounced granulocytopenia, thrombocytopenia, and a weakened immune system. There is a comprehensive decline in hematopoietic cells, particularly within the bone marrow. The subsequent phase of bone marrow failure is characterised by an inability to produce blood cells, crucial for maintaining immune status (Rödel et al., 2015). Consequently, the disorders induced by ionised cells include aplastic anemia, hypoplastic myelodysplastic syndrome, red blood cell aplasia, megakaryocytic thrombocytopenia, and chronic neutropenia (Lover, 1989).

Radiation exposure can result in immunosuppression and immunomodulation, which have different but interconnected impacts on the immune system. Radiation can weaken the immune system by damaging immune cells' DNA, harming bone marrow, and impairing T-cell development and function by damaging thymus and lymphoid tissues. It can also generate reactive oxygen species that weaken immune cells (Lover, 1989). Conversely, immunomodulation alters the immune response by affecting cytokine production and release, which is essential for immune regulation. Radiation can trigger an inflammatory response by releasing substances like TNF-α and interleukins, potentially enhancing certain immune functions. Furthermore, it can influence regulatory T cells (T-reg), which are vital for maintaining immune tolerance and preventing autoimmunity, and alter the adaptive immune response. Irradiation affects the expression levels of immune- receptors and the cytotoxic response in peripheral blood mononuclear cells (Ozturk et al.,2023; Jenkins et al. 2024).

2.3.6. Acute Radiation Syndrome

Acute Radiation Syndrome (ARS) is a severe illness resulting from the exposure of a substantial portion of the body to penetrating radiation over 0.7 Gy over a brief period (Garau et al., 2011). This exposure can lead to immediate and acute effects (Acosta & Warrington, 2020). The gravity of ARS is contingent upon the radiation dose, with higher doses inducing more severe symptoms and, without treatment, possibly causing death within days or weeks. Symptoms of ARS typically manifest within hours to days after exposure and may include nausea, vomiting, diarrhea, fever, weakness, fatigue, and changes in blood cell counts. After a period of well-being during which the symptoms recede, the distinct disease phase begins with disease-specific symptoms. ARS impacts the hematopoietic, gastrointestinal, and cardiac/neurological systems, with each system requiring a different dose to manifest symptoms (Shin et al., 2020). ARS persists for several months after exposure. For those who can recover, the healing process can take anywhere from a few weeks to years (Acosta & Warrington, 2020).

3. Stochastic Effects

Stochastic effects are of particular concern in the context of low-dose radiation exposure, such as that experienced in medical imaging procedures or occupational settings. The two main types of stochastic effects are eğigenetic and hereditary effects.

3.1. Epigenetic Effects

Exposure to radiation causes oxidative stress responses and epigenetic changes in normal tissues (Shin et al., 2020). When exposed to radiation, cells immediately produce free radicals with a nanosecond half-life. The redox system generates free radicals several hours after exposure, which may continue for years. The carcinogenic effects of radiation are due to the ability of ionising radiation to damage DNA within cells, leading to mutations that can result in uncontrolled cell growth and tumour formation.

Radiation can cause oxidative stress to biological systems directly or indirectly through its byproducts. These effects may be localised or systemic. Additionally, ionising radiation can induce non-targeted or multi-targeted effects (Morgan & Sowa, 2015). Various oxidases produced by the redox system and free radicals contribute to tissue damage through different regulatory mechanisms (Spitz et al., 2004). The significance of the redox system in both the initial and subsequent stages of radiation effects has been increasingly recognized (Yahyapour et al., 2018). Reactive oxygen and nitrogen species are the primary culprits of radiation normal tissue effects. Irradiation-induced free radicals are known to activate various enzymes, such as nicotinamide adenine dinucleotide phosphate oxidase, lipoxygenases, nitric oxide synthase, and cyclooxygenases, which are differentially expressed across a range of cells, tissues, and organs (Wei et al., 2019). These enzymes contribute to the radiation effect pathway via diverse regulatory mechanisms. Epigenetic modulators are crucial in the biological radiation effects process.

The primary epigenetic modifications include the regulation of non-coding ribonucleic acid (RNA), histone modifications, and DNA methylation (Zhou et al., 2020). MicroRNAs and non-coding RNAs are crucial regulators of numerous cancer characteristics, interacting with DNA, RNA, proteins, or encoding small peptides. Reactive oxygen species are recognized for regulating each phase of tumourigenesis by acting as secondary messengers within cancer cells. An imbalance in ROS leads to "oxidative stress," a pathological condition critical in regulating cancer progression. Furthermore, the interaction between oxidative stress and non-coding RNAs has been observed to govern the expression of many genes and the activation of various signaling pathways related to cancer hallmarks, indicating a potential mechanistic connection between ncRNAs, oxidative stress and cancer (Talapko et al., 2024).

Simultaneously, the induction of oxidative stress and increased levels of ROS within mitochondria can harm mitochondrial DNA. These mutations can affect the epigenetic regulatory mechanisms of nuclear DNA by diminishing the activity of methyltransferases, resulting in extensive DNA hypomethylation (Szumiel, 2015). The progeny of the irradiated cells inherits such changes. Chronic oxidative stress is a significant factor in delayed post-radiation outcomes, including cancer development.

Sporadic and radiation-induced carcinogenesis is characterized by the activation of oncogenes and the inactivation of tumor suppressor genes. These genes can be influenced by both genetic and epigenetic mechanisms. Over time, radiation can induce genomic instability, propelled by reactive oxygen species and cytokine cycles, leading to delayed genetic and epigenetic alterations. For most individuals exposed to radiation, the absolute risk of developing a malignancy induced by radiation is significantly lower than that of sporadic malignancies. In essence, radiation increases the incidence of malignancy in proportion to the dose and the age of the individual.

3.2. Hereditary Effects

Radiation exposure can cause mutations in the DNA of germ cells, leading to changes in the genetic information passed on to offspring. These mutations may result in genetic disorders, developmental abnormalities, or other health problems in future generations (Shayenthiran Sreetharan et al., 2023). The risk of hereditary effects is generally considered to be much lower than the risk of cancer and is typically of more significant concern at higher doses of radiation (Degenhardt et al., 2024).

Exposure to a high dose of radiation typically leads to permanent azoospermia, jeopardizing men's fertility due to decreased testosterone levels and sperm quality (Wdowiak et al., 2019). Consequently, the likelihood of fathering children diminishes. Moreover, adverse reproductive outcomes, such as stillbirths and congenital malformations, have been reported among radiation workers (Parker et al., 1999). Similar detrimental effects on germ cells have been observed in animal models subjected to radiation (Dhakal et al., 2021). It has been noted that fertility reduction due to parental radiation exposure can be transmitted to subsequent generations through germ cells (Baulch et al., 2007; Mughal et al., 2012). It was recorded that epigenetic changes, including DNA methylation changes, have been detected in transgerative and multi-beloved irradiation studies containing flora and fauna (Laanen et al., 2021). However, low-dose radiation given to male parents, which triggers a significant cytogenetic adaptive response in their germ cells, is unlikely to cause harmful effects on their offspring (Cai & Wang, 1995).

The evidence is insufficient to determine whether parental preconception radiation exposure is associated with health effects. This uncertainty largely stems from the heterogeneity of findings (Draper et al., 1997; Stephens et al., 2024). The absence of definitive evidence does not imply that such effects do not exist; instead, it suggests that if adverse health outcomes do occur in the offspring of exposed parents, measuring these effects consistently and reliably is challenging (Draper et al., 1997; Stephens et al., 2024). The inherent inconsistencies in research designs underscore a potential area of concern.

While the probability of stochastic effects occurring at low doses may be small for an individual, the cumulative risk across a population can be significant, highlighting the importance of minimizing unnecessary radiation exposure and implementing appropriate radiation protection measures. Stochastic effects are a significant concern with low doses over an extended period. Unlike deterministic effects, which increase in severity with the radiation dose and have a clear threshold below which they do not occur, stochastic effects can happen at any dose. However, the probability of occurrence increases with higher doses. The absence of a threshold for stochastic effects underlines the importance of minimizing radiation exposure to the lowest feasible levels, adhering to the ALARA (As Low As Reasonably Achievable) principle. This radiation effect is particularly significant in public health, radiological protection, and regulatory frameworks for mitigating long-term cancer risks and genetic mutations in exposed populations. Understanding the temporal pattern of radiation effects is essential for assessing risks, guiding medical management, and developing radiation protection and mitigation strategies.

Conclusion

The biological effects of radiation are significant and varied, mainly due to the ionization of molecules within cells, leading to direct DNA damage. This damage can result in single and double-strand breaks, base modifications, and cross-linking, which may cause cell death, cancer development, and inheritable genetic mutations. Acute exposure to high radiation doses can lead to immediate health issues like radiation sickness. In contrast, chronic exposure to lower doses may increase the risk of long-term effects such as cancer and cardiovascular and genetic impacts on subsequent generations. Understanding these effects is essential for creating protective measures and therapeutic medical, environmental, and occupational interventions. Therefore, the hazards associated with radiation exposure are not limited to immediate health concerns but also pose substantial long-term risks to individual and public health. Comprehending these mechanisms is crucial for developing protective and therapeutic strategies, highlighting the importance of radiation safety in various contexts.

References

Acosta R., Warrington S. J. *Radiation Syndrome.* [Updated 2023 Jan 16]. In: StatPearls [Internet]. Treasure Island (FL): StatPearls Publishing; 2024 Jan-. Available from: https://www.ncbi.nlm.nih.gov/books/NBK441931.

Afshari, N., Koturbash, I., Boerma, M., Newhauser, W., Kratz, M., Willey, J., Williams J., Chancellor J. (2024). A Review of Numerical Models of Radiation Injury and

Repair Considering Subcellular Targets and the Extracellular Microenvironment. *International Journal of Molecular Sciences*, *25*(2), 1015-15.

Amrenova, A., Baudin, C., Ostroumova, E., Stephens, J., Anderson, R., & Laurier, D. (2024). Intergenerational effects of ionizing radiation: review of recent studies from human data (2018–2021). *International Journal of Radiation Biology*, 1-11. A. https://doi.org/10.1080/09553002. 2024.2309917.

Ban, N., Clero, E., Vaillant, L., Zhang, W., Hamada, N., Preston, D., & Laurier, D. (2022). Radiation detriment calculation methodology: Summary of ICRP Publication 152. *Journal of Radiological Protection*, *42*(2), 023001.

Baulch, J. E., Li, M. W., & Raabe, O. G. (2007). Effect of ATM heterozygosity on heritable DNA damage in mice following paternal F0 germline irradiation. *Mutation Research/Fundamental and Molecular Mechanisms of Mutagenesis*, *616*(1-2), 34-45.

Cai, L., & Wang, P. (1995). Induction Of A Cytogenetic Adaptive Response In Germ Cells Of Irradiated Mice With Very Low-Dose Rate Of Chronic Γ-Irradiation And Its Biological Influence On Radiation-Induced DNA Or Chromosomal Damage And Cell Killing In Their Male Offspring. *Mutagenesis*, *10*(2), 95-100. 10.2.95.

Cooper, Griselda, and Robert E. Hausman. (2000). *The Cell: A Molecular Approach.* 2nd ed. Oxford: Sinauer Associates. gov/books/NBK9839/.

Crompton, A., Gamage, K., Jenkins, A. and Taylor, C. (2018). Alpha Particle Detection Using Alpha-Induced Air Radioluminescence: A Review and Future Prospects for Preliminary Radiological Characterisation for Nuclear Facilities Decommissioning. *Sensors,* [online] 18(4), p.1015. doi:https://doi.org/10.3390/s18041015.

Culligan, Kevin M., Clare E. Robertson, Foreman J., Doerner P., and Britt A B. (2006). "ATR and ATM Play Both Distinct and Additive Roles in Response to Ionizing Radiation." *The Plant Journal* 48 (6): 947–61. https://doi.org/10.1111/j.1365-313x.2006.02931.x.

Degenhardt, Ä., Sreetharan, S., Amrenova, A., Adam-Guillermin, C., Dekkers, F., Dumit, S., ... & Applegate, K. E. (2024). The ICRP, MELODI, and ALLIANCE Workshop on Effects of Ionizing Radiation Exposure in Offspring and Next Generations: A Summary of Discussions. *International Journal of Radiation Biology,* January, 1–11. https://doi.org/10. 1080/09553002 .2024.2306335.

Dhakal, R., Yosofvand, M., Yavari, M., Abdulrahman, R., Schurr, R., Moustaid-Moussa, N., & Moussa, H. (2021). Review Of Biological Effects Of Acute And Chronic Radiation Exposure On Caenorhabditis Elegans. *Cells*, *10*(8), 1966. https://doi.org/10.3390/ cells10081966.

Donya, M., Radford, M., ElGuindy, A., Firmin, D., & Yacoub, M. H. (2015). Radiation In Medicine : Origins, Risks, And Aspirations. *Global Cardiology Science and Practice,* 2014(4), 57. https://doi.org/ 10.5339/gcsp.2014.57.

Draper, G. J., Little, M. P., Sorahan, T., Kinlen, L. J., Bunch, K. J., Conquest, A. J., Kendall GM, Kneale GW, Lancashire RJ, Muirhead C., O'Connor CM, & Vincent, T. J. (1997). Cancer in the offspring of radiation workers: a record linkage study. *BMJ,* 315(7117), 1181-1188. PMID : 9393219 ; PMCID: PMC2127770.

Einor, D., Bonisoli-Alquati, A., Costantini, D., Mousseau, T. A., & Møller, A. P. (2016). Ionizing Radiation, Antioxidant Response And Oxidative Damage: A Meta-

Analysis. *Science of the Total Environment*, 548, 463-471. https://doi.org/10.1016/j.scitotenv.2016.01.027.

Hada, M., Wu, H., & Cucinotta, F. A. (2011). mBAND Analysis For High-And Low-LET Radiation-Induced Chromosome Aberrations: A Review. *Mutation Research/ Fundamental and Molecular Mechanisms of Mutagenesis*, 711(1-2), 187-192. 12.018.

Hall, Eric J, and Giaccia A J. (2019). *Radiobiology for the Radiologist.* 8th ed. Philadelphia: Wolters Kluwer.

Hendry, J. H. (2012). "Radiation Biology and Radiation Protection." *Annals of the ICRP* 41 (3-4): pp. 64–71. https://doi.org/10.1016/j.icrp.2012.06.013.

Henry, E., and Arcangeli M-L. (2020). "How Hematopoietic Stem Cells Respond to Irradiation: Similarities and Differences between Low versus High Doses of Ionizing Radiation." *Experimental Hematology,* December. https://doi.org/10.1016/j.exphem.2020.12.001.

Huang, Y., Zhang, W., Yu, F., & Gao, F. (2017). The cellular and molecular mechanism of radiation-induced lung injury. *Medical science monitor: international medical journal of experimental and clinical research*, pp. 23, 3446–50. https://doi.org/10.12659/msm.902353.

i Garau, M. M., Calduch, A. L., & López, E. C. (2011). Radiobiology of the Acute Radiation Syndrome. *Reports of Practical Oncology and Radiotherapy*, 16(4), 123-130. https://doi.org/10.1016/j.rpor.2011.06.001.

Jayan, J., Roshi, H., Ashraf, F. F. P., Nair, P. G., Vijayakumar, A., Nair, A. S., Pappachen L. K., Abdelgawad M. A., Parambi D. G. T., Aleya L., Mathew, B. (2022). Effects of radiation exposure on brain health: a state of the art and new challenges. *Environmental Science and Pollution Research*, 29(58), 87068–87081. Doi: 10.1007/s11356-022-23703-4. Epub 2022 Oct 29. PMID: 36308656.

Jenkins, S. V., Johnsrud, A. J., Dings, R. P. M., & Griffin, R. J. (2024). Bystander Effects in Spatially Fractionated Radiation Therapy: From Molecule To Organism To Clinical Implications. *Seminars in radiation oncology*, 34(3), 284–291.

Jung, H., Beck-Bornholdt, H. P., Svoboda, V., Alberti, W., & Herrmann, T. (2001). Quantification Of Late Complications After Radiation Therapy. *Radiotherapy and Oncology,* 61(3), 233-246. https://doi.org/ 10.1016/s0167-8140(01)00457-1.

Klammer, H., Mladenov, E., Li, F., & Iliakis, G. (2015). Bystander Effects as Manifestation of Intercellular Communication of DNA Damage and of the Cellular Oxidative Status. *Cancer letters,* 356(1), 58-71. https://doi.org/10.1016/j.canlet.2013.12.017.

Laanen, P., Saenen, E., Mysara, M., Van de Walle, J., Van Hees, M., Nauts, R., Van Nieuwerburgh Fi., ... & Horemans, N. (2021). Changes In DNA Methylation In Arabidopsis Thaliana Plants Exposed Over Multiple Generations To Gamma Radiation. *Frontiers in Plant Science,* 12, 611783. https://doi.org/10.3389/fpls.2021.611783.

Little, M P. (2009). "Cancer and Non-Cancer Effects in Japanese Atomic Bomb Survivors." *Journal of Radiological Protection* 29 (2A): A43–59. .

Little, M. P., Azizova, T. V., Bazyka, D., Bouffler, S. D., Cardis, E., Chekin, S., Chumak Vadim V.,... & Lipshultz, S. E. (2012). Systematic Review and Meta-Analysis of

Circulatory Disease from Exposure to Low-Level Ionizing Radiation And Estimates of Potential Population Mortality Risks. *Environmental health perspectives,* 120(11), 1503-1511. https://doi.org/10.1289/ehp.1204982.

Lopes, J., Leuraud, K., Klokov, D., Durand, C., Bernier, M. O., & Baudin, C. (2022). Risk Of Developing Non-Cancerous Central Nervous System Diseases Due To Ionizing Radiation Exposure During Adulthood: Systematic review and meta-analyses. *Brain Sciences,* 12(8), 984. https://doi.org/10.3390/brainsci12080984.

Lover, J. (1989). Radiobiological Properties Of Human Hematopoietic And Stromal Marrow Cells. *Stem Cells,* 7(4), 203–212. https://doi.org/10.1002/stem.5530070402.

Lu, Q., Liang, Y., Tian, S., Jin, J., Zhao, Y., & Fan, H. (2023). Radiation-Induced Intestinal Injury: Injury Mechanism And Potential Treatment Strategies. *Toxics,* 11(12),1011-11. https://doi.org/10.3390/toxics11121011.

MacNaughton W. K. (2000). Review article: new insights into the pathogenesis of radiation-induced intestinal dysfunction. *Alimentary pharmacology & therapeutics,* 14(5), 523–528. https://doi.org/10.1046/j.1365-2036.2000.00745.x.

Marechal, A., and Zou. L. (2013). "DNA Damage Sensing by the ATM and ATR Kinases." Cold Spring Harbor, *Perspectives in Biology* 5 (9): a012716–16. https://doi.org/
10.1101/cshperspect.a012716.

Mehta, K. (2005). "Radiation: Basic Principles." *Journal of Vascular Surgery* 42 (6): 1237–38. https://doi.org/10.1016/j.jvs.2005.08.022.

Morgan, W F. (2011). "Radiation-induced genomic instability." *Health Physics* 100 (3): 280–81. https://doi.org/10.1097/hp.0b013e3182082f12.

Morgan, William F., and Sowa M. B. (2015). "Non-Targeted Effects Induced by Ionizing Radiation: Mechanisms and Potential Impact on Radiation-Induced Health Effects." *Cancer Letters* 356 (1) : 17–21. https://doi.org/10.1016/j. canlet .2013.09.009.

Mughal, S. K., Myazin, A. E., Zhavoronkov, L. P., Rubanovich, A. V., & Dubrova, Y. E. (2012). The Dose And Dose-Rate Effects Of Paternal Irradiation On Transgenerational Instability In Mice: A Radiotherapy Connection. Edited by Gayle E. Woloschak. *PLoS ONE* 7 (7): e41300. https://doi.org/10.1371/journal.pone.0041300.

Nikitaki, Z., Nikolov, V., Mavragani, I. V., Mladenov, E., Mangelis, A., Laskaratou, D. A., Fragkoulis G. I., & Georgakilas, A. G. (2016). Measurement of complex DNA damage induction and repair in human cellular systems after exposure to ionizing radiations of varying linear energy transfer (LET). *Free radical research,* 50(sup1), S64-S78. https://doi.org/10.1080/10715762.2016. 1232484.

Olivares, A., Alcaraz-Saura, M., Achel, D. G., Berná-Mestre, J. D. D., & Alcaraz, M. (2021). Radiation-induced bystander effect: Loss of radioprotective capacity of rosmarinic acid in vivo and in vitro. *Antioxidants,* 10(2), 231-31. https://doi.org/10.3390/antiox10020231.

Ozasa, K., Shimizu, Y., Suyama, A., Kasagi, F., Soda, M., Grant, E. J., Sakata R., Sugiyama H., and Kodama, K. (2012). Studies of the Mortality of Atomic Bomb Survivors, Report 14, 1950–2003: An Overview of Cancer and Noncancer Diseases. *Radiation research,* 177(3), 229-243. https://doi.org/10.1667/rr2629.1.

Ozturk, N., Karlıtepe, A., Depboylu, B., & Eren, M. K. (2023). Immunomodulatory effects of ionizing radiation on peripheral blood mononuclear cells. *International Journal of Radiation Research*, 21(1), 73-78.

Parker, L., Pearce, M. S., Dickinson, H. O., Aitkin, M., & Craft, A. W. (1999). Stillbirths among offspring of male radiation workers at Sellafield nuclear reprocessing plant. *The Lancet,* 354(9188), 1407–1414. https://doi.org/ 10.1016/s0140- 6736(99)04138-0.

Peters, C. E., Quinn, E. K., Rodriguez-Villamizar, L. A., MacDonald, H., & Villeneuve, P. J. (2023). Exposure To Low-Dose Radiation In Occupational Settings And Ischaemic Heart Disease: A Systematic Review And Meta-Analysis. *Occupational and Environmental Medicine,* 80(12), 706-714. https://doi.org/10.1136/oemed-2023-108865.

Prise, K. M., Pinto, M., Newman, H. C., & Michael, B. D. (2001). A Review Of Studies Of Ionizing Radiation-Induced Double-Strand Break Clustering. *Radiation Research,* 156(5), 572–576. https://doi.org/10.1667/ 0033-7587(2001)156[0572:arosoi]2.0.co;2.

Ramos-Méndez, J., LaVerne, J. A., Domínguez-Kondo, N., Milligan, J., Štěpán, V., Stefanová, K., Perrot Y... & Faddegon, B. (2021). TOPAS-nBio validation for simulating water radiolysis and DNA damage under low-LET irradiation. *Physics in Medicine & Biology,* 66(17), 175026-26. https://doi.org/10.1088/1361-6560/ac1f39.

Rödel, F., Frey, B., Multhoff, G., & Gaipl, U. (2015). Contribution Of The Immune System To Bystander And Non-Targeted Effects Of Ionizing Radiation. *Cancer Letters,* 356(1), 105-113. https://doi.org/10.1016/j.canlet. 2013.09.015.

Shadad, A. K., Sullivan, F. J., Martin, J. D., & Egan, L. J. (2013). Gastrointestinal Radiation Injury: Symptoms, Risk Factors And Mechanisms. *World Journal Of Gastroenterology: WJG,* 19(2), 185. https://doi. org /10.3748/wjg.v19.i2.185.

Shin, E., Lee, S., Kang, H., Kim, J., Kim, K., Youn, H., Jin Y. W., Seo S.,& Youn, B. (2020). Organ-Specific Effects Of Low Dose Radiation Exposure: A Comprehensive Review. *Frontiers in genetics,* 11, 566244. https://doi.org/10.3389/fgene.2020. 566244.

Sia, J., Szmyd, R., Hau, E., & Gee, H. E. (2020). Molecular Mechanisms Of Radiation-Induced Cancer Cell Death: A Primer. *Frontiers in cell and developmental biology,* 8, 41. https://doi.org/10.3389/fcell.2020.00041.

Sies, H. (2015). Oxidative Stress: A Concept in Redox Biology and Medicine. *Redox Biology* 4 (April): pp.180–83. https://doi.org/10.1016/j.redox. 2015.01.002.

Spitz, D. R., Azzam, E. I., Jian Li, J., & Gius, D. (2004). Metabolic Oxidation/Reduction Reactions and Cellular Responses to Ionizing Radiation: A Unifying Concept in Stress Response Biology. *Cancer and Metastasis Reviews,* 23 (3/4): 311-322. https://doi.org/10.1023/b:canc. 0000031769.14728.bc.

Sreetharan, S., Frelon, S., Horemans, N., Laloi, P., Salomaa, S., & Adam-Guillermin, C. (2023). Ionizing Radiation Exposure Effects Across Multiple Generations: Evidence And Lessons From Non-Human Biota. *International Journal of Radiation Biology,* 1-18. https://doi.org/10.1080/09553002. 2023.2281512.

Srivastava, T., Chirikova, E., Birk, S., Xiong, F., Benzouak, T., Liu, J. Y., Villeneuve P.J.,& Zablotska, L. B. (2023). Exposure To Ionizing Radiation And Risk of

Dementia: A Systematic Review and Meta-Analysis. *Radiation Research,* 199(5), 490-505. https://doi.org/10.1667/rade-22-00153.1.

Stephens, J., Moorhouse, A. J., Craenen, K., Schroeder, E., Drenos, F., & Anderson, R. (2024). A Systematic Review Of Human Evidence For The Intergenerational Effects Of Exposure To Ionizing Radiation. *International Journal of Radiation Biology,* 1-34. https://doi.org/10.1080/09553002 .2024.2306328.

Szumiel, I. (2015). Ionizing Radiation-Induced Oxidative Stress, Epigenetic Changes and Genomic Instability: The Pivotal Role of Mitochondria. *International Journal of Radiation Biology,* 91(1), 1–12. https://doi.org/10.3109/09553002.2014.934929.

Talapko, J., Talapko, D., Katalinić, D., Kotris, I., Erić, I., Belić, D., Mihaljević MV., ... & Škrlec, I. (2024). Health Effects of Ionizing Radiation on the Human Body. *Medicina,* 60(4), 653-53. https://doi.org/10.3390/medicina 60040653.

Tapio, S. (2016). Pathology and Biology of Radiation-Induced Cardiac Disease. *Journal of Radiation Research* 57 (5): 439–48. https://doi.org/10.1093/ jrr/rrw064.

Tapio, S., Little, M. P., Kaiser, J. C., Impens, N., Hamada, N., Georgakilas, A. G., Simar. & Salomaa, S. (2021). Ionizing Radiation-Induced Circulatory and Metabolic Diseases. *Environment International,* 146, 106235. https://doi.org /10.1016/j.envint. 2020.106235.

Tohidinezhad, F., Di Perri, D., Zegers, C. M., Dijkstra, J., Anten, M., Dekker, A., Van Elmpt W., Eekers D. B. P.& Traverso, A. (2022). Prediction Models For Radiation-Induced Neurocognitive Decline In Adult Patients With Primary Or Secondary Brain Tumors: A Systematic Review. *Frontiers in Psychology,* 13, 853472. https://doi.org/ 10.3389/fpsyg.2022.853472.

Vetter, R. J. (2008). ICRP Publication 103, the Recommendations of the International Commission on Radiological Protection. *Health Physics* 95, no. 4 (October 2008) : 445–46. https://doi.org/10.1097/01.hp.0000324200. 73903.5b.

Wardman, P. (2009). The Importance of Radiation Chemistry to Radiation and Free Radical Biology (the 2008 Silvanus Thompson Memorial Lecture). *The British Journal of Radiology* 82 (974): 89–104. https://doi.org/ 10.1259/bjr/60186130.

Wdowiak, A., Skrzypek, M., Stec, M., & Panasiuk, L. (2019). Effect of ionizing radiation on the male reproductive system. *Annals of Agricultural and Environmental Medicine,* 26(2). 210–16. .https://doi.org/10. 26444/ aaem/106085.

Wei, J., Wang, B., Wang, H., Meng, L., Zhao, Q., Li, X., Xin Y. & Jiang, X. (2019). Radiation-Induced Normal Tissue Damage: Oxidative Stress And Epigenetic Mechanisms. *Oxidative medicine and cellular longevity,* 2019(1), 3010342. https://doi.org/10.1155/ 2019/3010342.

Yahyapour, R., Motevaseli, E., Rezaeyan, A., Abdollahi, H., Farhood, B., Cheki, M., ... & Villa, V. (2018). Reduction–Oxidation (Redox) System In Radiation-Induced Normal Tissue Injury: Molecular Mechanisms And Implications In Radiation Therapeutics. *Clinical and Translational Oncology,* 20(8), 975-988. https://doi.org/10.1007/s12094-017-1828-6.

Yu, Z., Xu, C., Song, B., Zhang, S., Chen, C., Li, C., & Zhang, S. (2023). Tissue Fibrosis Induced By Radiotherapy: Current Understanding Of The Molecular Mechanisms, Diagnosis And Therapeutic Advances. *Journal of Translational Medicine,* 21(1), 708. https://doi.org/10.1186/s12967-023-04554-0.

Zhou, L., Zhang, Z., Huang, Z., Nice, E., Zou, B., & Huang, C. (2020). Revisiting Cancer Hallmarks: insights from the interplay between oxidative stress and non-coding RNAs. *Molecular Biomedicine,* pp. 1, 1–24. https://doi.org/10.1186/s43556-020-00004-1.

Chapter 4

Effects of Radiation on the Foetus

Bengü Depboylu[1,2,*]
[1]Department of Radiation Oncology, Aydın Adnan Menderes University, Aydın, Turkiye
[2]Department of Basic Oncology and Cancer Biology, Health Sciences Institute, Aydın Adnan Menderes University, Aydın, Turkiye

Abstract

Ionising radiation has teratogenic, mutagenic, and carcinogenic effects on a developing embryo and foetus. The consequences of in-utero radiation exposure are strongly correlated with the gestational week of the pregnancy in which radiation exposure occurred, the dose and dose distribution of the radiation, and environmental factors. Congenital malformations, intellectual disability, microcephaly, seizures, growth retardation, increased cancer risk, and foetal death are attributed to the deterministic effects of radiation which need a threshold dose of <0.1Gy. Stochastic effects are not dose-related, and carcinogenesis is a typical example. All female patients of childbearing age should be questioned (e.g., last menstrual period) for potential pregnancy prior to imaging interventions. A pregnancy test should be run before proceeding if needed. The ideal time for elective diagnostic intervention is the first ten days of the menstrual cycle to skip an unacknowledged pregnancy. Cancer in pregnant women is uncommon, with an estimated incidence of 0.02 to 0.1% and affecting 1 in 1000-2000 pregnancies. Pregnancy itself is not a predisposing condition for cancer.
Melanoma, breast, cervix, ovary, thyroid cancers, lymphoma and leukaemia are the most common types of cancer in pregnant women. Oncologic management of the pregnant should be multidisciplinary, and decisions should be made by weighing the pros and cons of the

* Corresponding Author's Email: bengu.depboylu@adu.edu.tr.

In: Triage in Radiation Accidents
Editors: Nural Öztürk and Sibel Karaca
ISBN: 979-8-89530-415-0

treatment modalities for maternal and foetal sides. Termination of pregnancy should be discussed with the patient in case of advanced-stage disease or aggressive pathologies if malignancy is diagnosed in early pregnancy. The two conditions parental radiation exposure could affect the next generation are in-utero foetal exposure and preconceptional gonadal exposure. While atomic bomb survivors do not exhibit any heritable effects on the next generation, animal and epidemiological studies contradict each other. To answer the need for proper study outcome assessment and accurate exposure/dose-effect relationship interpretation, reliable dosimetry, precise reporting, and result reproducibility are crucial. This section focuses on the effect of ionising radiation on the foetus.

Keywords: foetus, in-utero radiation exposure, ionising radiation, pelvic irradiation, prenatal irradiation

1. Introduction

Regarding the potential health risks of ionising radiation on humankind, in-utero exposure has always been a specific point of interest for researchers due to the unique sensitivity of embryonal and foetal tissues. As early as the end of the 1920s, the disturbances in the neonates of women who received pelvic irradiation were reported (Goldstein and Murphy, 1929); however, the physical and biological data were partly missing. A decade later, these disturbances were comprehensively evaluated in animal embryos, including radiation type, dose, and gestational age (Job, Leibold, and Fitzmaurice,1935). However, concluding these effects in humans is impossible because of the differences between the species and selection bias.

Recently, a deeper understanding of the biological effects of low-dose ionising radiation has increased concerns. The atomic bomb survivors of World War II (WWII) who were exposed to radiation in utero demonstrated mental retardation, microcephaly, seizures, and intellectual disabilities in their later life compared to their healthy peers (Ishihara et al., 2022; Schull and Otake, 1986). Despite the limited data (small sample size and significant uncertainties), their condition was attributed to the radiation-induced damage of the developing nervous system in utero. Foetuses exposed to >0.5 Gy doses between 8-15 weeks after conception have suffered neurocognitive defects (Otake and Schull, 1998; Schull and Otake, 1999).

Moreover, a higher incidence of leukaemia in the 0-24 age group near nuclear installations in England and Wales has been documented (Forman et al., 1987; Goldsmith, 1992). Following the Chornobyl hazard in 1986, supporting increases in leukaemia were reported in Greece, Germany, Scotland, and Belarus (Gibson et al., 1988; Kellerer,1998; Petridou et al., 1996; Steiner et al., 1998; Gibson et al., 1988; Kellerer, 1998).

Although the United Nations Scientific Committee on Effects of Atomic Radiation (UNSCEAR) denied the adverse outcomes in pregnant women due to the Chornobyl hazard in 2001, two decades later, there is enough evidence for the correlation of increased thyroid cancers in children born to ^{131}I internalised pregnant women (Busby et al., 2009; UNSCEAR, 2006). Additionally, children had dose-dependent, more extended gestational periods, microcephaly, and chest anomalies but average birth weights. No biological mechanism has yet to clarify the reason behind the more extended gestational period (UNSCEAR, 2006).

Embryogenesis in humans is a process that begins in the preimplantation of a fertilised ovum, evolves into an embryo, and then shifts to a foetus with functioning organs with varying susceptibility to external forces (Khan, 2023). Both ionising and non-ionising radiation have effects, mostly when deleterious ones are ionising radiation born. In the general population, the radiation dose to which the developing foetus is exposed depends on the maternal environment. Background radiation during a regular pregnancy is 0.75 to 1mSv, varying on the geographic location, aviation, and occupational conditions (Bottollier-Depois et al., 2000; Yoon and Slesinger, 2020).

The purpose of this chapter is to review the current information on the effects of radiation on the foetus and to present data that can be beneficial in evaluating possible ionising radiation-born risks.

2. Effects of Prenatal Radiation on the Foetus: Basic Facts

The primary effects of radiation on human conception are miscarriage, embryonic miscarriage, foetal or neonatal death, malformations, growth retardation, postnatal functional impairment, and carcinogenesis. The embryonic tissues have an effective regeneration capacity to repair the deleterious effects of radiation and overcome the damage, and normal growth is also possible as a pregnancy outcome (Michel, 1989). The consequences of in-utero radiation exposure are strongly correlated with the gestational week of the pregnancy in which radiation exposure occurred, the dose and dose distribution of the radiation, and environmental factors.

Radiation exerts its effects on the embryo or foetus in two ways: deterministic or stochastic.

Deterministic effects occur over a minimum threshold, which follows the "cause and effect" relationship. The probability of the specific impact shows a linear increase with the radiation dose. This threshold dose is described as low as 0.1Gy, for a range between 0.05-0.1Gy; the risks become indefinite and negligible at doses below 0.05Gy (Martin et al., 2009). Congenital malformations, intellectual disability, microcephaly, seizures, growth retardation, increased cancer risk, and foetal death are attributed to the deterministic effects of radiation (Brent, 2015). The developmental stages of organogenesis are the critical stage for pathologic findings of deterministic effects that become visible in miscarriage/stillborn foetuses due to exposure of a large number of cells to radiation (Yoshimaru et al., 1995).

Stochastic effects occur incidentally regardless of any threshold dose. The probability of specific shows a linear-quadratic relation with the dose. Increased risk of carcinogenesis is the result of stochastic effects. The increase in thyroid cancer in children of pregnant women who are exposed to radiation following the Chornobyl hazard is an example of stochastic effects (Busby et al., 2009).

Low-dose radiation exerts its damage on the embryo or foetus through radical oxygen species (ROS), thus going into reaction with significant biomolecules (DNA and RNA) to change their chemical structure, peroxidation of membrane phospholipids and oxidation of cellular proteins that regulate critical functions inside the cell (Andrés Juan et al., 2021). The regeneration capacity of the embryonic/foetal tissues plays a determining role at this point for the manifestation of the in-utero radiation-exposed organ system, such as lymphoproliferative tissues are more prone to these adverse effects than the neural tissue due to their rapid cell turnover (Mebius, 2007). It explains the increased infantile leukaemia cases near nuclear installations.

Ionising radiation has teratogenic, mutagenic, and carcinogenic effects on the developing embryo and foetus. The first two weeks after conception is the most delicate pregnancy period for the embryo's future because it may either be affected and resorbed or not affected and, consequently, continue to grow a foetus. The "all or none" phenomenon explains the course of events during this period (Adam, 2012). The embryo is still in the stem cell stage, and cleavage has not started yet. The effect of radiation on the embryo is determined by the number of dead or damaged cells, which lead to chromosomal damage or lethal malformations. When a certain number of cells are killed, the embryo withstands and usually develops. However, when the number of damaged or dead cells exceeds the average, the embryo is lost (Brent, 2008; 2015).

Animal studies showed that a small dose of 0.1Gy may kill the highly fragile developing embryo in the first two weeks, just before the implantation. Embryonic death may occur at tiny doses of 0.15 to 0.2 Gy in mice (Brent, 2009).

In the case of radiation exposure, whether exceptions of all or none phenomenon exist, further studies with animals revealed in the 1960s, high natural susceptibility of a mouse strain to a particular congenital anomaly, and in the 1980s, no increased risk for malformation in the preimplantation period was found below 0.2 Gy in a genetically susceptible strain (Pampfer and Streffer, 1988; Rugh, 1963).

Organogenesis starts in the second week postconception and continues till the end of the eighth week. Developing the embryo gains a radioresistant nature. Animal studies showed that a dose of 1Gy will harm the embryo, causing malformation (Chang et al., 2021). Radiation-induced cell damage may cause malformations (via disruption of cellular migration and proliferation or mitotic delay), particularly in the nervous system. The most frequent stigma of in-utero radiation exposure is microcephaly. Microcephaly is defined as a head circumference that is than two standard deviations (SD) below the mean for gender and age; when it is greater than three SD below the mean, it is called severe microcephaly (Abuelo, 2007; Stoler-Poria et al., 2010) Newborns of A-bomb survivors those are exposed before eight weeks of their pregnancy did not show severe intelligence quotient (IQ) (Otake and Schull 1993).

The literature data does not report a linear correlation between the severity of microcephaly and lower intelligence quotient (IQ) scores. Moreover, irrelevant of the head size, in the presence of mental retardation, the embryos are bound to have significantly lower IQ scores (Otake and Schull, 1993).

Malformations related to the organs during the embryonic period are defects of the neural tube (such as anencephaly, meningocele), coelomic structures (such as valvular malformations, intestinal malformations), pulmonary, circulatory, genitourinary organ systems, and limb buds. While neonates of the A-bomb survivors did not exhibit severe malformations, neonates of women who underwent therapeutic abdominal irradiation had bizarre malformations (Chang et al., 2021).

The foetal period extends from the ninth week until birth (the end of the 38th week). For any measurable effect, a dose of >0.1Gy is required. Irradiation during this stage reveals a small-sized foetus due to a diminishing number of developing foetal cells. Growth restriction is typically seen as low birth weight and short stature in adulthood. A proportioned permanent growth restriction was observed in the A-bomb survivors' follow-up, and the effect became more significant with an increase in radiation dose above 1Gy. Furthermore, with a cumulative exposure dose >1Gy in the first 12 weeks in-utero, a 3-4% short stature was reported upon reaching maturity (18 years) (Otake and Schull, 1993; 1998).

At 8 to 15 weeks, mental retardation becomes most explicit when an in-utero foetus exposes to radiation. While the neonate may exhibit medium to severe intellectual disability consequently, this incidence significantly diminishes with exposure beyond 25 weeks gestation. The risk is directly proportional to the exposed dose, and at 8-15 weeks, the threshold is 0.12Gy, and at 16-25 weeks, 0.21Gy (Nakashima 1994; Otake and Yoshimaru,1989; Otake and Schull,1998). Foetal doses >0.1Gy may result in some reduction of IQ (intelligence quotient).

Foetal doses in the range of 1Gy may lead to severe mental retardation and microcephaly, particularly during 8-15 weeks and, to a lesser extent, 16-25 weeks after conception (Neriishi, Monzen, and Okamoto, 2012). Correspondingly, the IQ loss is between 25-31 points with a 40% risk of developing severe intellectual disability per Gy above 0.1Gy. However, at 16-25 weeks, an average IQ loss drops to 13-21 points at doses >0.7 Gy, and severe intellectual disability risk drops to 9% per Gy at the same doses (Nakashima, 1994; Otake M, Yoshimaru H 1989; Otake and Schull 1998 Yoshimaru, 1989).

Carcinogenic effects of in-utero radiation exposure occur during the advanced stages of the foetal period. Ionising radiation interferes with DNA, causing damaged replication, increasing the frequency of natural mutations, and inducing carcinogenesis (Chang et al., 2021). Exposure to any dose is related to childhood malignancy, and the risk of leukaemia is increased almost twice with 0.01-0.02Gy doses (Kato, 1971; Mole, 1990). The excess childhood cancer risk is approximately 6% per Gy (Doll and Wakeford, 2000). A recent review stated that high exposure and short duration dampens the carcinogenic risk. While a protracted continuous radiation of 0.02 Gy/day does not increase the risk, an acute exposure of 1 Gy at mid-gestation fuels the risk. Irradiation of the late foetus is primarily due to the diagnostic medical interventions in the final weeks of pregnancy, and the risk of many childhood cancers is increased (Brent 2014). However, reaching reliable conclusions is difficult due to limited data and other issues (recall and selection bias) (Wakeford and Bithell, 2021).

A wide range of childhood solid cancers, excluding the bone tumours not typical for this period, are reported with an increased incidence due to antenatal exposure (Preston et al., 2008; Wakeford and Bithell, 2021). Even embryonic tumours like neuroblastoma and Wilm's tumour, which are confined to early childhood, are detected due to irradiation shortly before birth (Boice and Miller, 1999). Recent studies confirmed that leukaemia, thyroid cancer, and central nervous system tumours increased with exposure to moderate and high therapeutic doses in utero (Braganza et al., 2012).

3. Effects of Diagnostic Radiology Imaging During Pregnancy on the Foetus

All female patients of childbearing age should be questioned (e.g., last menstrual period) for potential pregnancy prior to imaging interventions. (Brent, 2009; Applegate, 2007). Under doubtful circumstances, a pregnancy test should be run before proceeding. Diagnostic imaging of pregnant women is an integral part of their follow-up during their pregnancy. Since they are exposed to radiation throughout the procedure, the medical team (physicians, radiologists, and radiology technicians) must be aware of the medical rationale of the imaging and its deleterious effects on the mother and foetus. The pregnant women should be informed in detail about the aim of the diagnostic imaging radiation-related potential harms to the foetus, and written consent should be obtained at every step of the imaging. The ideal time for elective diagnostic intervention is the first ten days of the menstrual cycle to skip an unacknowledged pregnancy; nevertheless, it is not 100% safe.

According to the report of The American Society of Radiologic Technologists, the total cumulative dose has risen six times over the last 45 years in conjunction with the increasing rate of computed tomography use (Hricak et al., 2011). Typically, foetal radiation doses exposed by diagnostic imaging are below 0.01Gy. In fluoroscopic interventions, a barium enema may exceed the threshold, yet by proper prerequisites, a dose reduction might be guaranteed (American College of Radiology, 2013). For a single

CT imaging, the foetal dose depends on variables, such as the scan's number and point of interest, and the slice thickness is between 0.01-0.04 Gy (Kruskal, 2022). Recommended techniques to reduce the foetal dose during CT scan are to modulate the CT tube, z-axis overscan, minimise the scan length, in short-statured and younger patients, modifying the kV used, use of iterative reconstruction algorithms and deep learning or artificial intelligence (Picone et al., 2023).

Upon requesting diagnostic imaging on pregnant women, the physician and the radiologist should take full responsibility and apply all necessary techniques for minimising the foetus from excess exposure. A radiologist consultation before the imaging would provide detailed maternofoetal medical data regarding the pending intervention, recheck maternal consent, and a medical physicist should calculate the dose to safeguard the foetus.

Providing the pregnant with a lead apron during a non-abdominopelvic intervention minimises the foetal dose to a significant extent. Digital radiograms may be used to minimise cumulative dose. Diagnostic imaging of other body parts like the head and neck, thorax, and extremities does not cause significant scatter radiation, which will harm the foetus. Physicians and radiologists must decide, according to the pregnant patient's best interest, to expose her and the foetus to radiation by weighing the pros and cons of diagnostic imaging (Kim and Boyd, 2022).

For abdominopelvic diagnostic imaging, a posteroanterior exposure is recommended to reduce the dose to 0.02-0.04 Gy since the foetus will be in the anterior pelvis. Furthermore, scattered radiation may be reduced via shuttered collimators. Grid usage, avoidance of magnification around the womb and repeated imaging should be applied.

For angiographic and fluoroscopic imaging interventions, the techniques to reduce the foetal dose are minimising the exposure duration, number of point-of-interest images as well as the area of intervention, and altering the beam size (Kruskal, 2022).

One advantage of scintigraphic imaging interventions is adequate hydration of the pregnant woman, and oft urination deliberately eliminates the radiation. Moreover, radionuclides with short half-lives are preferred (technetium-99m or I^{123}) to keep <0.01Gy exposure (American College of Obstetricians and Gynecologists, 2018). Though these points can benefit the foetus, limited studies about diagnostic imaging with radionuclides imply otherwise.

Interventions like thyroid scans should not be scheduled for pregnancy as the foetal thyroid will readily absorb the radioactive iodine, increasing the concerns for childhood thyroid cancer (De Santis et al., 2007). The literature data for positron emission tomography (PET) scans in pregnant women is scarce and primarily based on case series, which makes it difficult to estimate the results for both the pregnant and the foetus (Zanotti-Fregonara, 2012). Studies showed that foetal dose from PET/CT depends on foetal weight, gestational stage, administered maternal dose and radiotracer used (Zanotti-Fregonara et al., 2016; Xie and Zaidi, 2016; 2014). Diagnostic imaging with MRI or CT scans should be preferred in pregnancy instead of PET/CT, considering that they provide similar information. A technique to minimise the foetal dose, empty mother bladder, and more extended image acquisition period is advised when the pregnant woman's clinical decision of PET/CT scan is scheduled (Ahmadi, 2022).

Regarding sentinel lymph node biopsy (SLNB) and lymphoscintigraphy interventions in pregnant women, the literature data contains ambiguous studies since the 2000s. During the SLNB procedure, the ipsilateral breast receives 0.0022Gy, the contralateral breast 0.0009Gy and the abdomen 0.00045 Gy dose, and all below the accepted threshold when radioactive colloid technetium-99m is used as tracer (Gentilini et al., 2004). A decade later, studies regarding the safety of other pharmaceutical agents showed that methylene blue is also safe as technetium-99m (Gropper et al., 2014). A very recent study sealed the safety of SLNB; however, it declared blue dye as contraindicated and recommended technetium-99m (Konishi et al., 2022). It is worth highlighting that in half of the published articles in the literature, there is no proof for the exact dying method that creates its own bias over final results (Bothou et al., 2024).

4. Cancer During Pregnancy and Foetal Implications

Cancer in pregnant women is uncommon, with an estimated incidence of 0.02 to 0.1% and affecting 1in 1000-2000 pregnancies (de Haan et al., 2018; Pentheroudakis and Pavlidis, 2006; Voulgaris, Pentheroudakis, and Pavlidis, 2011). Contemporary women postpone pregnancy to their 30s and 40s; consequently, increasing age brings cancer and pregnancy doublet forth. More specialists are confronted by complex once-obstetrics decision-making (Wolters et al., 2021). Even though cancer rarely affects the growing foetus, it becomes more challenging to diagnose and treat. Pregnancy itself is not a predisposing factor to develop cancer, but cancer-related symptoms may mimic the pregnancy signs (e.g., gynecologic malignancies), therefore, may delay the diagnosis of cancer (Korenaga and Tewari, 2020; Van Calsteren et al., 2010).

Melanoma, breast, cervix, ovary, thyroid cancers, lymphoma and leukaemia are the most common types of cancer in pregnant women with decreasing order (Pavlidis, 2002; Skrzypczyk-Ostaszewicz and Rubach, 2016). Lung and gastric cancers have the worst prognosis with a coexisting pregnancy (Maggen et al., 2020; Rothschild, 2016).

Oncologic management during pregnancy should be multidisciplinary. Per the general principles, the oncologic treatment approach does not differ from an average patient; however, in a woman who wishes to keep her pregnancy, the medical team should decide by weighing the pros and cons of the management on maternal and foetal sides. Termination of pregnancy should be discussed with the patient in case of advanced-stage disease or aggressive pathologies if malignancy is diagnosed in early pregnancy. However, preterm delivery is risky for long-term morbidities of a premature neonate and, therefore, is not recommended (Lejarraga, 2016).

Management for diagnostic imaging and staging is discussed in detail in previous parts of this chapter. Options for treatment modalities to oncologic pregnancies should be considered according to potential adverse effects on maternal and foetal health. Surgery can be safely performed in any trimester whenever indicated. To prevent spontaneous abortions, the best time to perform a surgical intervention is between 8 and 15 weeks (Peccatori et al., 2013).

Systemic treatment options are chemotherapy, endocrine therapy, molecular targeted agents and immunotherapy. Most chemotherapy drugs are small light molecules with high permeability at the placental wall. They are also highly teratogenic. The timing of chemotherapy is essential as its deleterious effect highly correlates with gestational age. During the second and third trimesters, the physiological changes due to the pregnancy significantly change the chemotherapeutics' distribution area. After 35 weeks of pregnancy, discontinuation of chemotherapy is advised to allow adequate recovery of maternal and foetal bone marrows for the delivery (Esposito et al., 2016; Peccatori et al., 2013).

The natural course of a healthy pregnancy and foetal development is highly hormone dependent. Therefore, any manipulation that disrupts the hormonal balance will result in a foetal anomaly. Consequently, endocrine therapy during pregnancy is contraindicated (Poggio et al., 2020).

The placental passage of agents used for targeted therapy and immunotherapy depends on their structure and size. Large molecules (e.g., trastuzumab, rituximab) are actively transported from the 14th week, and small molecules (e.g., tyrosine kinase inhibitors) readily pass through the placenta anytime (Abruzzese et al., 2014; Lambertini, Peccatori, and Azim, 2015).

Definitive radiotherapy delivers 10^4 to 10^5 times higher doses to pregnant than diagnostic imaging (The Royal College of Radiologists, 2020). Radiotherapy is not recommended during pregnancy, considering its deleterious effects on the foetus and the neonate, but postponed until after delivery. In palliative settings, radiotherapy may be planned during pregnancy only after a thorough judgement of its benefits for the maternal health and safeguarding of the foetus (Chen et al., 2008). The foetal exposure dose should be calculated via a phantom model by the medical physicist, and appropriate modifications to the treatment plan should be made regarding field size, angle, and energy of the beam (Wolters et al., 2021).

During irradiation, three sources are recognised that are essential for the evaluation of the foetal dose: head leakage (D_{ex}), scatter from collimators and beam modifiers, and scatter generated inside the maternal body (D_{in}) (Mazonakis et al., 2008; van der Giessen, 1994). In supradiaphragmatic irradiations, lead shielding should be utilised on the pregnant abdomen and pelvis to minimise foetal exposure. However, in subdiaphragmatic irradiations, the foetus is directly inside the radiation field, and therapeutic abortus should be proposed. The D_{in} data before the radiotherapy enables radiation oncologists and medical physicists to know the foetal dose (if $<$ 0.01 Gy) and the necessity to utilise lead shielding (Kourinou et al., 2015).

5. Fundamental Facts of Preconceptional Radiation Exposure

There are two conditions in which parental radiation exposure could affect the next generation: in-utero foetal exposure and preconceptional gonadal exposure (Sadetzki and Flint-Richter, 2006). A-bomb survivors who were exposed to radiation and followed up in the postwar years failed to exhibit any heritable effects on the next generation (Nakamura, 2006). Interestingly, animal studies and epidemiological studies contradict each other. Despite an interesting case-control study from the UK in the late 1990s, epidemiological studies have not proven an association between parental gonadal irradiation and hereditary effects (Draper et al., 1997).

Radiobiology studies are another source of understanding the possible hereditary effects of exposed parents. For a proper assessment of study outcomes and an accurate understanding of the exposure/dose-effect relationship, reliable dosimetry, precise reporting, and result reproducibility are essential. While dose-related variables like radiation source, irradiation angles, and dose rate are typically well-known in radiobiology studies, crucial details of the irradiation process are frequently overlooked in reporting (Degenhardt, Dumit, and Giussani, 2023). Preserving fertility before undergoing oncologic treatment is delicate, depending on the medical team's expertise, the time requirement of a potential technique, and the patient's choice, which is affected by geographic, cultural, and economic difficulties (Emirdar et al., 2023).

Women are strongly advised to postpone potential pregnancy until after several months ahead of finishing the radiation therapy. Animal studies proved that oocyte maturity is the most crucial determiner of its radiosensitivity; the more mature the oocyte, the more radiosensitive it is. When the preconception ovarian dose is >0.5Gy, a delay of the pregnancy to 8 weeks and beyond is advised (Valentin, 2000).

The irradiation field in the body is closely related to the severity of the potential fertility and obstetric challenges of the patient. Irradiation to the cranium may rummage the hypothalamic hypophyseal axis, resulting in hormonal imbalances. Dysfunctional uterine bleeding due to hormonal imbalances is a significant problem in the fertility of young women who plan pregnancy (Hill-Kayser et al., 2023). Irradiation for childhood cancers to the craniospinal axis, total body (TBI), involved field, or whole pelvis increases the developmental abnormalities in bony structures, organ vasculature and texture. Uterine elasticity, distensibility, and resilience may compromise decidualisation (Critchley et al., 2002). Pelvic irradiation-related premature ovarian insufficiency is seen at doses >10Gy (TBI) to 30Gy (whole pelvis) at 90-97%, respectively, leading to decreased fertility window; therefore, such patients are the best candidates for fertility preservation techniques (Kelsey et al. 2022) A flank irradiation for Wilms tumour may open the door to foetal malposition, maternal hypertension, preterm delivery (Green et al., 2010).

6. Peripartum-Intrapartum and Postpartum Radiation Exposure to the Offspring

In pregnancy-associated cancers, the maternal state may require monitorisation of the foetus; thus, intrapartum radiation exposure becomes a great concern. The general principle denies any intervention with high doses of ionising radiation for diagnostic or treatment methods on behalf of the foetus (e.g., CT, MRI, fluoroscopy or angiography, scintigraphy, radiotherapy) (Valentin, 2000). Monitoring doses at birth and estimating exposures are necessary to balance gestational age and recommended delivery time (Parpinel et al., 2022). At term pregnancy, foetal exposure to diagnostic radiation has been described in the literature; PET/CT or whole-body diffusion-weighted/MRI is reported to be safely used just before the labour (Drouet et al., 2020; Parpinel et al., 2022; Drouet et al., 2020, Vandecaveye et al., 2021; Zanotti-Fregonara et al., 2022).

Intrapartum foetal exposure to radiation can be prevented by the safe use of ultrasound, which is pragmatically advised for foetal diagnostic imaging (Molina and Nicolaides, 2010).

Premature infants are more prone to radiation-induced deleterious effects. The portable X-rays are the most frequently used diagnostic imaging tools in neonatal intensive care units (NICU). The mean radiation exposure may increase to 3.65 mGy, and the mean scattered dose is 67.9 μGy (Cakir et al., 2021). The postnatal radiation exposure in these newborns is underestimated; it is essential to raise awareness regarding associated risks and strict criteria for requesting radiological exams (Toscan et al., 2024).

Conclusion

Ionising radiation has deterministic and stochastic effects on the growing foetus. Deterministic effects are dose-related (<0.1Gy) and follow a "cause and effect" relationship. Congenital malformations, intellectual disability, microcephaly, seizures, growth retardation, increased cancer risk, and foetal death occur in this way. Stochastic effects are not dose-related, and carcinogenesis is a typical example.

Parental radiation exposure is thought to affect the next generation with in-utero foetal exposure and preconceptional gonadal exposure. The consequences of in-utero radiation exposure are strongly correlated with the gestational week of the pregnancy in which radiation exposure occurred, the dose and dose distribution of the radiation, and environmental factors. A-bomb survivors with preconceptional gonadal exposure did not show heritable effects of radiation on the next generation. Epidemiologic and animal studies have conflicting results about the heritable effects of radiation. All women of childbearing age should be inquired for a possible pregnancy for diagnostic and/or therapeutic radiation exposure. A pregnancy test is advised when necessary for an unacknowledged pregnancy.

Cancer in pregnant women is uncommon, and its incidence is 1/1000-2000 pregnancies. Pregnancy-associated cancer is a unique condition that needs a multidisciplinary approach. Treatment decisions should be made by weighing the pros and cons of the treatment modalities for maternal and foetal sides.

References

Abruzzese, E., Trawinska, M. M., Perrotti, A. P., & De Fabritiis, P. (2014). "Tyrosine Kinase Inhibitors and Pregnancy." *Mediterranean Journal of Hematology and Infectious Diseases* 6 (1).

Abuelo, Dianne. (2007). "Microcephaly Syndromes." *Seminars in Pediatric Neurology* 14 (3): 118–27. https://doi.org/10.1016/j.spen.2007.07.003.

Adam, Margaret P. (2012). "The All-or-None Phenomenon Revisited." *Birth Defects Research Part A - Clinical and Molecular Teratology* 94 (8): 664–669.

Ahmadi, Amirhossein. (2022). "Radioprotective Agents in Radiotherapy-Induced Cell and Tissues Toxicity." *Anti-Cancer Agents in Medicinal Chemistry* 22 (1): 3–3.

Amant, F., Vandenbroucke, T., Verheecke, M., Fumagalli, M., Halaska, M. J., Boere, I., Han, S., Gziri, M. M., Peccatori, F., Rob, L., Lok, C., Witteveen, P., Voigt, J. U., Naulaers, G., Vallaeys, L., Van den Heuvel, F., Lagae, L., Mertens, L., Claes, L., Van Calsteren, K., International Network on Cancer, Infertility, and Pregnancy (INCIP). (2016). "Pediatric Outcome after Maternal Cancer Diagnosed during Pregnancy." *Archivos Argentinos de Pediatria* 114 (2): e121–22.

American College of Obstetricians and Gynecologists. (2018). "ACOG Committee Opinion." *Wolters Kluwer Health* 130 (3): 150–52.

American College of Radiology. (2013). "ACR- SPR Practice Parameter for Imaging Pregnant or Potentially Pregnant Adolescents and Women with Ionizing Radiation." American College of Radiology. 2013.

Applegate, Kimberly. (2007). "Pregnancy Screening of Adolescents and Women Before Radiologic Testing: Does Radiology Need a National Guideline?" *Journal of the American College of Radiology* 4 (8): 533–36.

Boice, J. D., Jr, & Miller, R. W. (1999). "Childhood and Adult Cancer after Intrauterine Exposure to Ionizing Radiation." *Teratology* 59 (4): 227–233.

Bothou, A., Margioula-Siarkou, C., Petousis, S., Margioula-Siarkou, G., Zervoudis, S., Sotiriadis, A., Amant, F., & Dinas, K. (2024). "Sentinel Lymph Node Biopsy for Breast Cancer during Pregnancy: A Comprehensive Update." *European Journal of Clinical Investigation* 54 (3): 1–13.

Bottollier-Depois, J. F., Quang, C., Bouisset, P., Kerlau, G., Plawinski, L.& Lebaron-Jacobs, L. (2000). "Assessing Exposure to Cosmic Radiation during Long-Haul Flights." *Radiation Research* 153 (5 I): 526–532.

Braganza, M. Z., Kitahara, C. M., Berrington de González, A., Inskip, P. D., Johnson, K. J., & Rajaraman, P. (2012). "Ionising Radiation and the Risk of Brain and Central Nervous System Tumors: A Systematic Review." *Neuro-Oncology* 14 (11): 1316–1324.

Brent, Robert L. (2008). "What Are the Reproductive and Developmental Risks of Ionising Radiation?" in *Reproductive and Developmental Risks of Ionizing Radiation* by Brent, Robert L. 56–58. Accessed May 19, 2024. https://core.ac.uk/download/pdf/46973281.pdf.

Brent, Robert L. (2009). "Saving Lives and Changing Family Histories: Appropriate Counselling of Pregnant Women and Men and Women of Reproductive Age, Concerning the Risk of Diagnostic Radiation Exposures during and before Pregnancy." *American Journal of Obstetrics and Gynecology* 200 (1): 4–24.

Brent, Robert L. (2014). "Carcinogenic Risks of Prenatal Ionising Radiation." *Seminars in Foetal and Neonatal Medicine* 19 (3): 203–213.

Brent, Robert L. (2015). *Protection of the Gametes Embryo/Foetus from Prenatal Radiation Exposure. Health Physics*. Vol. 108.

Busby, C., Lengfelder, E., Pflugbeil, S., & Schmitz-Feuerhake, I. (2009). "The Evidence of Radiation Effects in Embryos and Foetuses Exposed to Chernobyl Fallout and the Question of Dose Response." *Medicine, Conflict, and Survival* 25 (1): 20–40.

Cakir, S. C., Dorum, B. A., Koksal, N., Ozkan, H., Yazici, Z., Parlak, M., &Gulleroglu, N. B. (2021). "Radiation Exposure in the Neonatal Intensive Care Unit in Newborns and Staff." *American Journal of Perinatology* 40 (10): 1106–1111.

Chang, David S., Foster D. Lasley, Indra J. Das, Marc S. Mendonca, and Joseph R. Dynlacht. (2021). "Radiation Effects in the Embryo and Fetus" in *Basic Radiotherapy Physics and Biology: Second Edition*. Edited by Chang, David S., Foster D. Lasley, Indra J. Das, Marc S. Mendonca, and Joseph R. Dynlacht. 349-353. Springer Nature Switzerland.

Chen, Z., King, W., Pearcey, R., Kerba, M., & Mackillop, W. J. (2008). "The Relationship between Waiting Time for Radiotherapy and Clinical Outcomes: A Systematic Review of the Literature." *Radiotherapy and Oncology* 87 (1): 3–16.

Critchley, H. O., Bath, L. E., & Wallace, W. H. (2002). "Radiation Damage to the Uterus - Review of the Effects of Treatment of Childhood Cancer." *Human Fertility* 5 (2): 61–66.

de Haan, J., Lok, C. A. R., Schutte, J. S., van Zuylen, L., & de Groot, C. J. M. (2018). "Cancer Related Maternal Mortality and Delay in Diagnosis and Treatment: A Case Series on 26 Cases." *BMC Pregnancy and Childbirth* 18 (1): 1–6.

de Santis, M., Cesari, E., Nobili, E., Straface, G., Cavaliere, A. F., & Caruso, A. (2007). "Radiation Effects on Development." *Birth Defects Research Part C - Embryo Today: Reviews* 81 (3): 177–182.

Degenhardt, Ä., Dumit, S., & Giussani, A. (2023). "Effects of Ionising Radiation Exposure in Offspring and Next Generations: Dosimetric Aspects and Uncertainties." *International Journal of Radiation Biology* 0 (0): 1–7.

Doll, R., Wakeford, R. (2000). "Carcinogenesis After Exposure to Ionizing Radiation UCSF Radiology." *Invest Rad* 35:74–77.

Draper, G. J., Little, M. P., Sorahan, T., Kinlen, L. J., Bunch, K. J., Conquest, A. J., Kendall, G. M., Kneale, G. W., Lancashire, R. J., Muirhead, C. R., O'Connor, C. M., & Vincent, T. J. (1997). "Cancer in the Offspring of Radiation Workers: A Record Linkage Study." *British Medical Journal* 315 (7117): 1181–1188.

Drouet, C., Vrigneaud, J. M., Desmoulins, I., & Cochet, A. (2020). "FDG PET/CT in a Pregnant Woman with Breast Cancer." *Clinical Nuclear Medicine* 45 (7): e339–41.

Emirdar, V., Karatasli, V., Acet, F., Okay, G., Gode, F., Karabulut, A., & Arslan, Ç. (2023). "Perspectives and Knowledge about Fertility Preservation Strategies among Female Cancer Patients in Turkey." *International Journal of Clinical Practice, 2023*,6193187.

Esposito, S., Tenconi, R., Preti, V., Groppali, E., & Principi, N. (2016). "Chemotherapy against Cancer during Pregnancy: A Systematic Review on Neonatal Outcomes." *Medicine (United States)* 95 (38).

Forman, D., Cook-Mozaffari, P., Darby, S., Davey, G., Stratton, I., Doll, R., & Pike, M. (1987). "Cancer near Nuclear Installations." *Nature* 329 (6139): 499–505.

Gentilini, O., Cremonesi, M., Trifirò, G., Ferrari, M., Baio, S. M., Caracciolo, M., Rossi, A., Smeets, A., Galimberti, V., Luini, A., Tosi, G., & Paganelli, G. (2004). "Safety of Sentinel Node Biopsy in Pregnant Patients with Breast Cancer." *Annals of Oncology* 15 (9): 1348–1351.

Gibson, B. E., Eden, O. B., Barrett, A., Stiller, C. A., & Draper, G. J. (1988). "Leukaemia in Young Children in Scotland." *The Lancet* 332 (8611): 630.

Giessen, Piet Hein van der. (1994). "Calculation and Measurement of the Dose at Points Outside the Primary Beam for Photon Energies of 6, 10, and 23 MV." *International Journal of Radiation Oncology, Biology, Physics* 30 (5): 1239–1246.

Goldsmith, John R. (1992). "Nuclear Installations and Childhood Cancer in the UK: Mortality and Incidence for 0-9-Year-Old Children, 1971-1980." *Science of the Total Environment, The* 127 (1–2): 13–35.

Goldstein, L., & Murphy, D.P. (1929). "Etiology of the Ill-Health in Children Born after Maternal Pelvic Irradiation." *Am. J. Roentgenol.* 22:322–31.

Green, D. M., Lange, J. M., Peabody, E. M., Grigorieva, N. N., Peterson, S. M., Kalapurakal, J. A., & Breslow, N. E. (2010). "Pregnancy Outcome after Treatment for Wilms Tumor: A Report from the National Wilms Tumor Long-Term Follow-up Study." *Journal of Clinical Oncology* 28 (17): 2824–2830.

Gropper, A. B., Calvillo, K. Z., Dominici, L., Troyan, S., Rhei, E., Economy, K. E., Tung, N. M., Schapira, L., Meisel, J. L., Partridge, A. H., & Mayer, E. L. (2014). "Sentinel Lymph Node Biopsy in Pregnant Women with Breast Cancer." *Annals of Surgical Oncology* 21 (8): 2506–2511.

Hill-Kayser, C., Yorke, E., Jackson, A., Gracia, C., Ginsberg, J., Keene, K., Ronckers, C., Metzger, M. L., Li, Z., Constine, L. S., & Hua, C. H. (2023). "Effects of Radiation Therapy on the Female Reproductive Tract in Childhood Cancer Survivors: A PENTEC Comprehensive Review." *International Journal of Radiation Oncology Biology Physics*. 119(2): 588–609.

Hricak, H., Brenner, D. J., Adelstein, S. J., Frush, D. P., Hall, E. J., Howell, R. W., McCollough, C. H., Mettler, F. A., Pearce, M. S., Suleiman, O. H., Thrall, J. H., &

Wagner, L. K. (2011). "Managing Radiation Use in Medical Imaging: A Multifaceted Challenge." *Radiology* 258 (3): 889–905.

Ishihara, K., Kato, N., Misumi, M., Kitamura, H., Hida, A., & Yamada, M. (2022). "Radiation Effects on Late-Life Neurocognitive Function in Childhood Atomic Bomb Survivors: A Radiation Effects Research Foundation Adult Health Study." *Radiation Research* 197 (4): 403–407.

Ivanov, E. P., Tolochko, G. V., Shuvaeva, L. P., Ivanov, V. E., Iaroshevich, R. F., Becker, S., Nekolla, E., & Kellerer, A. M. (1998). Infant leukemia in Belarus after the Chernobyl accident. *Radiation and Environmental Biophysics*, 37(1), 53–55.

Job, T. T., Leibold, G.J. & Fitzmaurice, H. A. (1935). "Biological Effects of Roentgen Rays. The Determination of Critical Periods in Mammalian Development with X-rays." *American Journal of Anatomy* 56 (1): 97–117.

Juan, C. A., Pérez de la Lastra, J. M., Plou, F. J., & Pérez-Lebeña, E. (2021). "Molecular Sciences: The Chemistry of Reactive Oxygen Species (ROS) Revisited: Outlining Their Role in Biological Macromolecules (DNA, Lipids and Proteins) and Induced Pathologies." *Int. J. Mol. Sci* 22:4642.

Kato, Hiroo. (1971). "Mortality in Children Exposed to the A-Bombs While in Utero, 1945-1969." *American Journal of Epidemiology* 93 (6): 435–442.

Kelsey, T. W., Hua, C. H., Wyatt, A., Indelicato, D., & Wallace, W. H. (2022). "A Predictive Model of the Effect of Therapeutic Radiation on the Human Ovary." *PLoS ONE* 17 (11 November): 1–15.

Khan YS, Ackerman KM. (2023). "Embryology, Week I." In *StatPearls (Internet).* Treasure Island (FL): Stat Pearls Publishing. https://www.ncbi.nlm.nih.gov/books/NBK554562/.

Kim, Eunhye, & Brenda Boyd. (2022). "Diagnostic Imaging of Pregnant Women and Foetuses: Literature Review." *Bioengineering* 9 (6): 1–9.

Konishi, T., Fujiogi, M., Shigemi, D., Nishioka, K., Matsui, H., Fushimi, K., Tanabe, M., Seto, Y., & Yasunaga, H. (2022). "Surgical and Obstetric Outcomes of Breast Cancer Surgery during Pregnancy: A Nationwide Database Study in Japan." *Breast Cancer Research and Treatment* 195 (3): 289–299.

Korenaga, Travis Riley K., &Krishnansu S. Tewari. (2020). "Gynecologic Cancer in Pregnancy." *Gynecologic Oncology* 157 (3): 799–809.

Kourinou, K. M., Mazonakis, M., Lyraraki, E., &Damilakis, J. (2015). "Photon-Beam Radiotherapy in Pregnant Patients: Can the Foetal Dose Be Limited to 10cGy or Less?" *Physica Medica* 31 (1): 85–91.

Kruskal, Johnathan. (2022). "Diagnostic Imaging in Pregnant and Nursing Patients." UpToDate. 2022. Accessed May 25, 2024. https://medilib.ir/uptodate/show/4791.

Lambertini, M., Peccatori, F. A., & Azim, H. A., Jr. (2015). "Targeted Agents for Cancer Treatment during Pregnancy." *Cancer Treatment Reviews* 41 (4): 301–309.

Maggen, C., Lok, C. A., Cardonick, E., van Gerwen, M., Ottevanger, P. B., Boere, I. A., Koskas, M., Halaska, M. J., Fruscio, R., Gziri, M. M., Witteveen, P. O., Van Calsteren, K., Amant, F., & International Network on Cancer, Infertility and Pregnancy (INCIP). (2020). Gastric cancer during pregnancy: A report on 13 cases and review of the literature with focus on chemotherapy during pregnancy. *Acta OAbstetricia et Gynecologica Scandinavica*, 99(1), 79–88.

Martin, C. J., Sutton, D. G., West, C. M., & Wright, E. G. (2009). "The Radiobiology/Radiation Protection Interface in Healthcare." *Journal of Radiological Protection* 29 (2).

Mazonakis, M., Zacharopoulou, F., Varveris, H., &Damilakis, J. (2008). "Peripheral Dose Measurements for 6 and 18 MV Photon Beams on a Linear Accelerator with Multileaf Collimator." *Medical Physics* 35 (10): 4396–4403.

Mebius, Reina E. (2007). "Lymphoid Organogenesis: Educating Stroma." *Immunology and Cell Biology* 85 (2): 79–80.

Michel, C. (1989). "Radiation Embryology." *Experientia* 45 (1): 69–77.

Mole, R. H. (1990). "Childhood Cancer after Prenatal Exposure to Diagnostic X-Ray Examinations in Britain." *British Journal of Cancer* 62 (1): 152–168.

Molina, Francisca S., &Kypros H. Nicolaides. (2010). "Ultrasound in Labor and Delivery." *Foetal Diagnosis and Therapy* 27 (2): 61–67.

Nakamura, Nori. (2006). "Genetic Effects of Radiation in Atomic-Bomb Survivors and Their Children: Past, Present and Future." *Journal of Radiation Research* 47 (SUPPL. B): 67–73.

Nakashima, E. (1994). "Relationship of Five Anthropometric Measurements at Age 18 to Radiation Dose among Atomic Bomb Survivors Exposed in Utero." *Radiation Research* 138 (1): 121–126.

Neriishi, K., Monzen, Y., & Okamoto, N. (2012). "Radiation Effects on Pregnant Women, Foetuses, and Children." *Nihon Rinsho. Japanese Journal of Clinical Medicine* 70 (3): 451–456.

Otake M, Yoshimaru H, Schull WJ. (1989). "Prenatal Exposure to Atomic Radiation and Brain Damage." *Congenital Abnormalities* 29:309–320.

Otake, M., & Schull, W.J. (1993). "Radiation-Related Small Head Sizes among Prenatally Exposed A-Bomb Survivors." *International Journal of Radiation Biology* 63 (2): 255–270.

Otake, M., & Schull, W.J. (1998). "Review: Radiation-Related Brain Damage and Growth Retardation among the Prenatally Exposed Atomic Bomb Survivors." *International Journal of Radiation Biology* 74 (2): 159–171.

Pampfer, S., &Streffer, C. (1988). "Prenatal Death and Malformations after Irradiation of Mouse Zygotes with Neutrons or X-rays." *Teratology* 37 (6): 599–607.

Parpinel, G., Laudani, M. E., Giunta, F. P., Germano, C., Zola, P., & Masturzo, B. (2022). "Use of Positron Emission Tomography for Pregnancy-Associated Cancer Assessment: A Review." *Journal of Clinical Medicine* 11 (13), 3820.

Pavlidis, Nicholas A. (2002). "Coexistence of Pregnancy and Malignancy." *The Oncologist* 7 (6): 573–573.

Peccatori F. A., Azim H. A. Jr., Orecchia R., Hoekstra H. J., Pavlidis N., Kesic V, Pentheroudakis G.; ESMO Guidelines Working Group. (2013). "Cancer, Pregnancy and Fertility: ESMO Clinical Practice Guidelines for Diagnosis, Treatment and Follow-Up." *Annals of Oncology* 24 (SUPPL.6): vi160–70.

Pentheroudakis, George,& Pavlidis, N. (2006). "Cancer and Pregnancy: Poena Magna, Not Anymore." *European Journal of Cancer* 42 (2): 126–140.

Petridou, E., Trichopoulos, D., Dessypris, N., Flytzani, V., Haidas, S., Kalmanti, M., Koliouskas, D., Kosmidis, H., Piperopoulou, F., & Tzortzatou, F. (1996). "Infant

Leukaemia after in Utero Exposure to Radiation from Chernobyl." *Nature* 382 (6589): 352–353.

Picone, C., Fusco, R., Tonerini, M., Fanni, S. C., Neri, E., Brunese, M. C., Grassi, R., Danti, G., Petrillo, A., Scaglione, M., Gandolfo, N., Giovagnoni, A., Barile, A., Miele, V., Granata, C., & Granata, V. (2023). "Dose Reduction Strategies for Pregnant Women in Emergency Settings." *Journal of Clinical Medicine* 12 (5).

Poggio, F., Tagliamento, M., Pirrone, C., Soldato, D., Conte, B., Molinelli, C., Cosso, M., Fregatti, P., Del Mastro, L., &Lambertini, M. (2020). "Update on the Management of Breast Cancer during Pregnancy." *Cancers* 12 (12): 1–17.

Preston, D. L., Cullings, H., Suyama, A., Funamoto, S., Nishi, N., Soda, M., Mabuchi, K., Kodama, K., Kasagi, F., & Shore, R. E. (2008). "Solid Cancer Incidence in Atomic Bomb Survivors Exposed in Utero or as Young Children." *Journal of the National Cancer Institute* 100 (6): 428–436.

Rothschild, Sacha I. (2016). "Lung Cancer in Pregnancy-A Forgotten Disease Entity." *Journal of Thoracic Oncology* 11 (9): 1376–1378.

Rugh, R. (1963). "The Impact of Ionizing Radiations on the Embryo and Foetus." *The American Journal of Roentgenology, Radium Therapy, and Nuclear Medicine* 89 (January): 182–190.

Sadetzki, Siegal, and Flint-Richter, P. (2006). "Transgenerational Effects of Parental Exposure to Ionizing Radiation." *Harefuah* 145 (7):516–521.

Schull, W. J., Otake, M. (1986). "Neurological Deficit among the Survivors Exposed in Utero to the Atomic Bombing of Hiroshima and Nagasaki: A Reassessment and New Directions" in *Radiation Risks to the Developing Nervous System.* Eds. Kriegel, H., Schmahl, W., Gerber, G. B., Stieve, F.-E. Ž. Gustav Fischer Verlag, Stuttgart, New York, 399–419.

Schull, William J., & Otake, M. (1999). "Cognitive Function and Prenatal Exposure to Ionizing Radiation." *Teratology* 59 (4): 222–226.

Skrzypczyk-Ostaszewicz, Anna, & Rubach, M. (2016). "Gynaecological Cancers Coexisting with Pregnancy - A Literature Review." *Wspolczesna Onkologia* 20 (3): 193–198.

Steiner, M., Burkart, W., Grosche, B., Kaletsch, U., & Michaelis, J. (1998). "Trends in Infant Leukaemia in West Germany in Relation to in Utero Exposure Due to the Chernobyl Accident." *Radiation and Environmental Biophysics* 37 (2): 87–93.

Stoler-Poria, S., Lev, D., Schweiger, A., Lerman-Sagie, T., & Malinger, G. (2010). "Developmental Outcome of Isolated Foetal Microcephaly." *Ultrasound in Obstetrics and Gynecology* 36 (2): 154–58.

The Royal College of Radiologists. (2020). "IR (ME)R Implications for Clinical Practice in Diagnostic Imaging, Interventional Radiology and Diagnostic Nuclear Medicine." *Clinical Radiology*, no. June:1–131. www.rcr.ac.uk.

Toscan, M., de Araújo, B. F., Martini, J. C., Ravazio, R., & de Souza, V. C. (2024). "Our Estimates of Neonatal Radiation Exposure Fall Short of Reality." *European Journal of Pediatrics* 183 (4): 1911–1916.

UNSCEAR. (2006). "Report to the General Assembly, with Scientific Annexes." *Effects of Ionizing Radiation.* 2006. Accessed May 14, 2024. https://www.unscear.org/unscear/en/publications/2006_1.html.

Valentin, Jack. (2000). "Publication 84 Pregnancy and Medical Radiation." *Annals of IRCP* 30 (1): 1–42.

Van Calsteren, K., Heyns, L., De Smet, F., Van Eycken, L., Gziri, M. M., Van Gemert, W., Halaska, M., Vergote, I., Ottevanger, N., & Amant, F. (2010). "Cancer during Pregnancy: An Analysis of 215 Patients Emphasising the Obstetrical and the Neonatal Outcomes." *Journal of Clinical Oncology* 28 (4): 683–689.

Vandecaveye, V., Amant, F., Lecouvet, F., Van Calsteren, K., & Dresen, R. C. (2021). "Imaging Modalities in Pregnant Cancer Patients." *International Journal of Gynecological Cancer* 31 (3): 423–431.

Voulgaris, E., Pentheroudakis, G., & Pavlidis, N. (2011). "Cancer and Pregnancy: A Comprehensive Review." *Surgical Oncology* 20 (4): e175–185.

Wakeford, Richard, & Bithell, J. F. (2021). "A Review of the Types of Childhood Cancer Associated with a Medical X-Ray Examination of the Pregnant Mother." *International Journal of Radiation Biology* 97 (5): 571–592.

Wolters, V., Heimovaara, J., Maggen, C., Cardonick, E., Boere, I., Lenaerts, L., & Amant, F. (2021). "Management of Pregnancy in Women with Cancer." *International Journal of Gynecological Cancer* 31 (3): 314–22.

Xie, Tianwu, & Zaidi, H. (2014). "Foetal and Maternal Absorbed Dose Estimates for Positron-Emitting Molecular Imaging Probes." *Journal of Nuclear Medicine* 55 (9): 1459–1466.

Xie, Tianwu, & Zaidi, H. (2016). "Development of Computational Pregnant Female and Foetus Models and Assessment of Radiation Dose from Positron-Emitting Tracers." *European Journal of Nuclear Medicine and Molecular Imaging* 43 (13): 2290–2300.

Yoon, Ilsup, and Todd L. Slesinger. (2020). *Radiation Exposure In Pregnancy. StatPearls*. Treasure Island (FL): StatPearls Publishing. http://www.ncbi.nlm.nih.gov/pubmed/31869154.

Yoshimaru, H., Otake, M., Schull, W. J., &Funamoto, S.(1995). "Further Observations on Abnormal Brain Development Caused by Prenatal A-Bomb Exposure to Ionizing Radiation." *International Journal of Radiation Biology* 67 (3): 359–371.

Zanotti-Fregonara, P., Chastan, M., Edet-Sanson, A., Ekmekcioglu, O., Erdogan, E. B., Hapdey, S., Hindie, E., & Stabin, M. G. (2016). "New Foetal Dose Estimates from 18F-FDG Administered during Pregnancy: Standardisation of Dose Calculations and Estimations with Voxel-Based Anthropomorphic Phantoms." *Journal of Nuclear Medicine* 57 (11): 1760–1763.

Zanotti-Fregonara, P., Ishiguro, T., Yoshihara, K., Ishii, S., & Enomoto, T. (2022). "18F-FDG Foetal Dosimetry Calculated with PET/MRI." *Journal of Nuclear Medicine* 63 (10): 1592–1597.

Zanotti-Fregonara, Paolo. (2012). "Pregnancy Should Not Rule out 18FDG PET/CT for Women with Cancer." *The Lancet* 379 (9830): 1948.

Chapter 5

Radiation Accidents: Causes and Prevention Methods

Emriye Algül*

Department of Radiation Oncology, Kartal Dr. Lütfi Kırdar City Hospital, İstanbul, Türkiye

Abstract

Historically, following the discovery of radiation, humanity has encountered numerous radiation accidents, often resulting in catastrophic outcomes. These incidents have predominantly led to severe and permanent detrimental effects on both human health and the environment. Historical events such as the Chernobyl disaster and the Fukushima Daiichi nuclear power plant accident have demonstrated the immense destructive consequences of radiation accidents. Radiation-related accidents typically appear as nuclear power plant accidents, medical radiation accidents or industrial radiation accidents. These accidents mostly occur due to human errors, technical failures, or natural disasters. To prevent radiation accidents, training personnel and ensuring the continuity of this training, creation and implementation of comprehensive safety protocols, integration of technological innovations into existing system, taking preventive measures against radiation accidents, including regular and effective inspections are among the methods of minimizing possible risks. Considering the negative effects of radiation accidents on human health and their destructive impact on the environment and ecosystem, the importance of radiation protection becomes much clearer. With developing technology and procedures, protection from radioactive substances and

* Corresponding Author's Email: c_emriye@hotmail.com.

In: Triage in Radiation Accidents
Editors: Nural Öztürk and Sibel Karaca
ISBN: 979-8-89530-415-0

radiation are becoming increasingly effective. In this section, the causes of radiation accidents, their types and methods of protection against radiation accidents are explained.

Keywords: causes of radiation accidents, radiation accidents, prevention methods, types of radiation accidents

1. Introduction

An accident is used to describe a situation that results in harmful consequences due to an undesirable and unexpected event. Radiation accidents refer to situations where the harmful effects of ionising radiation occur in an unintended and unexpected manner (Turai and Veress, 2001). The types of radiation accidents are generally defined by the areas where radiation is used, such as nuclear power plant accidents, medical radiation accidents, and industrial radiation accidents (Coeytaux et al., 2015). Today, radiation has a very wide range of applications, and its usage areas are becoming increasingly widespread. Both the proliferation of its use and the global increase in the transportation of radioactive materials, naturally lead to an increase in accidents. Therefore, measures to prevent radiation accidents should be enhanced and continuously developed. For methods developed to prevent radiation accidents, it is necessary to examine past accidents, analyse the errors that caused them, and establish safety protocols based on the obtained results (ATSDR, 1999). Preventing radiation accidents is a critical necessity for the safety of not only the current generation but also future generations.

2. What Are Radiation Accidents?

A radiation accident is an uninvented or unexpected event caused by errors during the application of ionising radiation, leading to serious consequences due to the exposure of people or the environment (Cerezo, 2011; Turai and Veress, 2001).

Radiation has a wide range of uses worldwide, from medicine to industry, research laboratories to the military. The increasing use of radioactive materials in these areas raises the possibility of radiation accidents in many fields (IAEA Pub1102, 2000).

Radiation accidents can occur in areas such as nuclear facilities, medical facilities, industrial areas, research laboratories, or during transportation. These accidents most commonly occur in radiation devices containing sealed sources like ^{60}Co, ^{192}Ir, ^{137}Cs, or in X-ray devices used for diagnostic, therapeutic, or testing purposes. Around one-third of radiation accidents occur in medical applications while the source of one-third is not clearly identified. Radiation accidents occurring during transportation are the least common (Cerezo, 2011; Turai and Veress, 2001).

Radiation accidents can vary widely in nature, severity, and consequences, ranging from minor incidents with limited effects to major disasters with significant health, environmental, and socioeconomic impacts. Factors such as the type of radiation, duration of exposure, type of exposure (internal, external, or both), scale of the emergency, and existence of protective measures significantly alter the effects of radiation (Cerezo, 2011).

In radiation accidents, individuals may be exposed to external or internal radiation. In external radiation, the radiation sources are entirely outside of the body, and ionising radiation is not deposited in the body. When radiation contacts the body, it undergoes various interactions depending on the physical properties of the rays and radiation absorbed by the body. In internal radiation, radioactive material penetrates the body through the skin, respiratory system and digestive system. In this case, the individual is exposed to ionising radiation types such as alpha, beta, gamma, or neutron through internal irradiation, often resulting in much more destructive effects (Saenger, 1986).

Radiation accidents can cause serious problems in terms of emergency responses, public health, environmental, and socioeconomic impacts. While emergency interventions and medical treatments following radiation accidents are of great importance, taking necessary precautions to prevent these accidents is even more crucial. Therefore, it is essential to thoroughly understand the potential causes of radiation accidents and take all possible preventive measures accordingly.

3. Types of Radiation Accidents

Radiation accidents can be categorised into various types based on their occurrence locations or severity. Below are the main types of radiation accidents according to their occurrence locations.

3.1. Nuclear Power Plant Accidents

Nuclear power plant accidents are the accidents that occur in nuclear power plants and other nuclear facilities. Technical malfunctions in the reactor, human errors, or natural disasters can lead to these accidents. The accidents at nuclear power plants can result in the release of large amounts of radioactive material into the atmosphere. Nuclear power plant accidents are the most severe in terms of intensity and have the greatest environmental impact among radiation accidents.

The largest nuclear power plant accidents to date are the Chernobyl (1986) and Fukushima Daiichi (2011) disasters. These accidents were classified as Level 7 on the International Nuclear and Radiological Event Scale (INES). In the Vladimir Ilyich Lenin Nuclear Power Plant (Chernobyl Nuclear Power Plant) accident, an unexpected power surge occurred during a test in Reactor 4, leading to a series of reactions despite pressing the emergency button. The nuclear core melted, and the resulting fire spread a radioactive cloud over a wide area (IAEA Report INSAG-7, 1992).

In the Fukushima Nuclear Power Plant accident, a 9.0 magnitude earthquake triggered a 14-metre-high tsunami, damaging the electrical grid. The plant's generators were flooded, causing a power outage and insufficient cooling. This led to partial meltdowns and explosions. Due to the risk of leakage, a 20 km radius area was evacuated, and many workers at the plant were exposed to excessive radiation (IAEA Non-serial Publications, 2015).

3.2. Medical Radiation Accidents

Medical radiation accidents are accidents related to radiation sources used in medical fields. They can occur due to misuse, miscalibration or technical failures of radiation sources used in hospitals and clinics. Incorrect dose applications during radiotherapy, accidentally applying the planned dose to a different patient or to a different area from the treatment area of the same patient can be given as examples of such accidents.

When radiation accidents from 1980 to 2013 were analysed in ten-year periods, an increase in medical radiation accidents was observed. Although there was an increase in radiotherapy and fluoroscopic accidents, it was reported that there was a decrease in fluoroscopic accidents between 2000 and 2009. In this analysis, it was recorded that the number of overdosed people was increasing (Burtt et al., 2023).

One of the serious medical radiation accidents occurred in Costa Rica in 1996 due to a calibration error. A medical physicist incorrectly calibrated the beam output after replacing the source in a ^{60}Co device, resulting in a 66% longer irradiation time. This error caused 114 patients to receive a considerable overdose in a month, leading to severe side effects and the radiation-related deaths of 13 patients within two years (Sowby, 1981).

3.3. Industrial Radiation Accidents

Industrial radiation accidents usually originate from devices used in industrial radiography, thickness measurements, or sterilisation processes. These accidents result from the misuse or lack of maintenance of radiography devices, resulting to radiation exposure for workers or the environment (Bomanji, Novruzov, and Vinjamuri, 2014).

On October 17, 2016, an industrial radiography accident occurred in Sakarya, Turkey. During a radiography operation at a dam construction site near Kızılcıklı village, the source fell out of the device without being noticed. The guide tube and the slide cable were removed, and the device was placed in a vehicle. On October 18, a visitor to the site picked up the source, put it in his back pocket, and travelled by public transport for 2.5 hours. He then took it home, exposing other family members to radiation. The accident occurred due to a malfunctioning lock system on the gammagraphy device and the failure to perform radiation measurements during storage and transportation (TAEK Teknik Rapor 2017/4, 2017).

3.4. Research Laboratory Radiation Accidents

Research laboratory accidents are accidents occurring during the use of radiation in scientific research. They usually result from non-compliance with laboratory safety protocols. An example is the American physicist working on the Manhattan Project who accidentally exposed himself to radiation on August 21, 1945, at the Los Alamos Laboratory in New Mexico. He died 25 days later due to the accident. He was building a neutron reflector that surrounded the plutonium core with tungsten carbide bricks, acting as a radiation shield. He accidentally dropped a tungsten carbide brick onto a 6.2 kg bomb core made of plutonium-gallium alloy, triggering a critical reaction. A 'blue glow' was observed and he proceeded to disassemble the critical material and its tamper. He received radiation dosage sufficient to result in injuries from which he died 24 days later (Hayes, 1956).

3.5. Transportation-Related Radiation Accidents

Transportation-related radiation accidents are accidents that occur during the transportation of radiation sources. As the use of radioactive sources has become more widespread, transportation-related radiation accidents have become more frequent. These accidents generally result from a lack of training, packaging errors, incorrect documentation, or damage during transportation. In 2012, 12 accidents and incidents related to the transportation of radioactive sources were reported in United Kingdom with the most common causes being incorrect documentation or inadequate worker training (Hughes and Harvey, 2006).

The 1998-1999 Istanbul Ikitelli radiation accident involves a serious transportation related. A transport company packaged three ^{60}Co sources, received approval from the relevant authorities, and stored them in their Ankara depot for a while. Later, they sent these sources to their branch office in Istanbul where the employees placed two transport packages in an empty warehouse. The warehouse was then sold to a new owner, who sold the packages to a scrap dealer. The scrap dealers opened the box, removed the outer metal casing, and they were exposed to the ^{60}Co source. Medical and hematological examinations were performed on 404 people potentially exposed to radiation, and 18 people including seven children were hospitalised. While two containers containing radioactive sources were transferred to Istanbul, the second source was never found. An IAEA report on this accident indicated that the transport company's and original supplier's records were unreliable; it was unclear whether there was a radioactive source in the second container. The investigations by the Turkish Atomic Energy Authority were ongoing (IAEA STI/PUB/1102, 2000).

3.6. Military and Security-Related Radiation Accidents

Military and security-related radiation accidents are the radiation accidents that occur during the use of nuclear weapons or security research for military purposes. The accidents occurring during nuclear weapon tests fall into this category. These accidents are usually kept secret. In 1957, a large area in the

Ural Mountains was contaminated with fission products released accidentally from a secret facility. However, this accident only became known in 1990. After the accident, 20 villages were evacuated (Nénot, 1998).

4. Scaling Radiation Accidents

The International Nuclear and Radiological Event Scale (INES), developed by the International Atomic Energy Agency (IAEA) and the Nuclear Energy Agency of the Organisation for Economic Co-operation and Development (OECD/NEA), was implemented in 1990 to classify the significance of nuclear and radiological events and facilitate information sharing about these events. Initially used to classify events at nuclear power plants, the scale has since been applied to incidents occurring during the transportation, storage, or use of radioactive materials. The use of INES not only enables a clear understanding of the severity of nuclear and radiological events by everyone but also facilitates rapid and comprehensible communication of the level of importance of such events, thus promoting international information sharing. Additionally, it promotes the use of a common language for the evaluation and reporting of incidents among different countries and organisations (IAEA-INES-2009, 2013).

The INES scale utilises an 8-level scale to classify events, ranging from 0 to 7 (Figure 1). This scale categorises events considering their impacts on human health, the environment, and radiological barriers.

According to the classification depicted in:

Level 0 (Deviation): Events of no significance for safety. The 2008 Krsko Nuclear Power Plant coolant leak.

Level 1 (Anomaly): In this level of accidents, low radiation doses are given to employees or the public or minor errors occur in safety systems. It may involve a minor exceedance of an individual's annual dose limit or low-level radiation exposure due to a malfunction in a medical imaging device. On January 11, 2024, in Indonesia, a gamma camera containing a high-activity Ir-192 source was lost during transportation. On January 13, 2024, the gamma camera was found. When the source was found, no damage to the equipment was observed, and no measurable increase in radiation was detected around the equipment. This accident was reported as a Level 1 transportation radiation accident based on investigations conducted on May 28, 2014 (IAEA News, 2024).

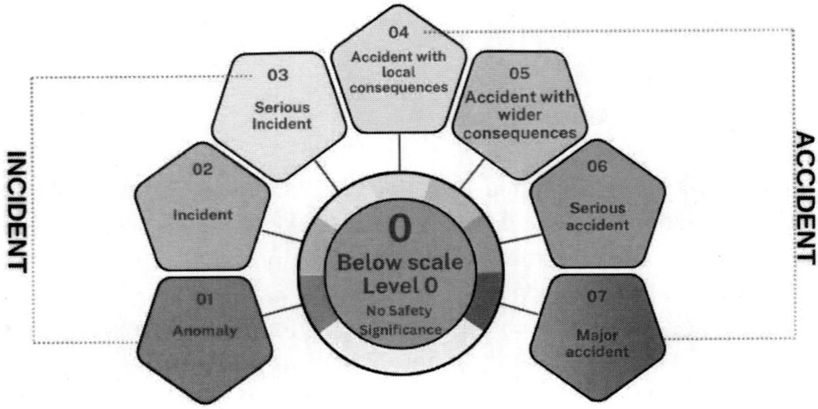

Figure 1. The International Nuclear and Radiological Event Scale.

Level 2 (Incident): It may be possible for an individual to exceed the annual dose limit. Examples include a radiotherapy patient receiving an excessive dose or the spread of low-level radioactive material to the environment. An example of a Level 2 incident can be cited in 2013 in France, where a practitioner in interventional radiology exceeded the annual dose limit due to overexposure (IAEA News, 2013).

Level 3 (Serious Incident): There is significant radiation exposure to employees or the public but it does not cause serious health effects. Examples include a security system failure at a plant resulting in high doses to workers or a radiologist receiving a high dose. In 2012, a radiologist's high dose exposure can be cited as an example of a Level 3 radiation incident (IAEA STI/PUB/1776, 2018).

Level 4 (Accidents with Local Consequences): In this level of radiation accidents, high doses of radiation may spread within the facility and employees may encounter serious health problems. In 2011, an accident in a gamma irradiation facility in Bulgaria resulted in four workers being exposed to excessively high doses of radiation classified as a Level 4 radiation accident according to the INES scoring (IAEA News, 2011).

Level 5 (Accidents with Wider Consequences): Involves the significant release of radiation outside a facility, resulting to high doses to the public and the environment. Limited public health effects may occur. Nuclear fuel melting and the release of radioactive materials, such as the

Three Mile Island accident in 1979 (Ahearne, 1980) is the example of Level 5.

Level 6 (Serious Accidents): There is significant release of very high levels of radiation outside a facility, causing in serious public health effects. It constitutes an accident requiring the evacuation of large areas due to the widespread release of radiation, such as the Kyshtym disaster in 1957 (IAEA STI/PUB/1644, 2014).

Level 7 (Major Accidents): It is the highest level on the Ines Scale. In accidents at this level, extremely high levels of radiation are released over a wide area. Serious and extensive health and environmental effects occur, as in the Chornobyl disaster in 1986.

5. Causes of Radiation Accidents

Although radiation accidents are relatively rare, they pose significant risks to both human health and environmental safety. Investigating and understanding the causes of these accidents is crucial for developing strategies to prevent future incidents and improving radiation safety protocols. Examination of past radiation accidents reveals numerous contributing factors. In the majority of these cases, human error has played a significant role (Holmberg, 2007). The following section delves into these causes in detail, providing a comprehensive analysis of the factors contributing to radiation accidents.

5.1. Human Errors

Errors by users working with radiation sources can lead to radiation accidents. Accidents often occur due to negligence, lack of communication, improper equipment use, lack of training, or failure to follow safety protocols. Examples of human errors include carelessness during device calibrations, improper protocol use, or incorrect use of dosimetric equipment, resulting in incorrect dose calculations and applications or improper use of radiation-emitting devices.

A study examining accidental high-dose irradiations in new radiotherapy technologies reported that human errors were the leading cause (62.5%) between 2000 and 2009 (Tsegmed et al., 2017).

In an incident in Panama between 2000 and 2001, after changes were made to the protocol used to enter blocks into the treatment planning system, verification was not performed, and 28 patients received high doses due to faulty calculations in the treatment planning system (IAEA, 2001).

In a 2007 incident in France, commissioning was carried out for the BrainLAB Novalis device installed in a clinic for stereotactic treatments. This device had a micro-MLC structure with a width of 3mm. For this device, capable of treating very small areas of 6x6 mm, parameters such as percent depth dose, beam profile, and scatter factors needed to be loaded into the treatment planning system. Incorrect measurement detectors were used by the physicist for these measurements. These incorrect measurement data were also entered into the treatment planning system. All patient plans were made with this faulty data. In this accident, 145 patients were affected (Derreumaux et al., 2008).

5.2. Technical Malfunctions

Radiation accidents can occur due to mechanical or electronic malfunctions in radiation-emitting devices. Such malfunctions are often attributed to inadequate maintenance and repair, the use of aged and worn-out equipment, or design flaws. Examples of accidents resulting from technical malfunctions including cooling system failures in nuclear reactors or the inability to retract radioactive sources due to mechanical problems in brachytherapy devices, resulting in the source being left exposed.

In a 2001 incident in Poland, during a power outage, a linear accelerator unexpectedly shut down during treatment. Subsequently, the treatment continued for five patients. However, two patients experienced itching and burning sensations during treatment, leading to treatment cessation and dosimetric measurements. These measurements revealed that the device's output was significantly higher than expected. Subsequent investigations into the cause of the error found that the accelerator's dose monitoring system was malfunctioning, and the safety interlock system's electronic components had failed. As a result of this technical malfunction, five patients were exposed to excessively high doses (Eichholz, 2004).

5.3. Natural Disasters

Floods, earthquakes, tsunamis, or other natural disasters can cause damaging to nuclear facilities or areas where radiation sources are present, leading to radiation leakage. A notable example is the Fukushima Daiichi nuclear power plant accident following a major earthquake and subsequent tsunami in Japan in 2011 (IAEA Non-serial Publications, 2015).

5.4. Sabotage and Terrorism

Intentional use of radiation in attacks aimed at sabotage and terrorism or the spread of radiation due to security vulnerabilities resulting from terrorist attacks on nuclear facilities can be responsible for radiation accidents. Especially following the September 11, 2001, attacks, enhanced security measures and international threats necessitated much effective protection against nuclear attacks or the use of radioactive materials for malicious purposes. In this context, all countries, whether they have nuclear power plants or not, need to consider this issue and establish national regulations regarding nuclear safety (IAEA, 2006).

5.5. Lack of Training

Incorrect practices that can cause accidents may occur due to insufficient knowledge of radiation safety or inadequate training regarding the use of radiation-producing devices. Ignorance or disregarding of radiation safety protocols can also lead to radiation accidents.

5.6. Management and Organisational Issues

Inadequate risk assessments, failure to conduct safety inspections, or failure to develop, implement emergency plans by management, and organisational structures can contribute to radiation accidents.

5.7. Faulty Design and Construction

Faulty design, construction of nuclear facilities or radiation-emitting devices can generate security vulnerabilities and radiation accidents.

Between 1985 and 1987, a radiation accident occurred with the Therac-25 computer-controlled linear accelerator device where patients were exposed to excessive doses of radiation. When the X-ray mode was selected, the process of setting and verifying the treatment device parameters took about 20 seconds. After the X-ray mode was selected and switched quickly to the electron mode, the failure of the X-ray target and filter to engage resulted in high-speed electrons produced in the X-ray mode being delivered directly to the patient, causing the accident. It happened six different times in hospitals. Right after the event was first reported, it was not accepted as an error by the manufacturer. Afterward, when the incident was reported in the other hospital, it was considered as a microswitch error. Even though more incidents are recorded, the manufacturer still did not accept it as an error. A month later, when the same accident occurred for the second time at the same hospital, it was identified as an error by the hospital physicist. It was announced that this accident was due to a production error. Six patients were affected by the failure, and the death of two patients was associated with excessive doses (IAEA STI/PUB/1084, 2000).

5.8. Monitoring and Control Deficiencies

Radiation levels in areas where radiation is used should be regularly monitored and controlled. Poor management of monitoring and control processes in these areas or neglecting them prevents the early detection of potential problems. In addition, failure to regularly calibrate radiation monitoring devices or the inadequate use of these devices may delay the detection of possible leaks and therefore lead to radiation accidents.

6. Methods of Preventing Radiation Accidents

Measures to prevent radiation accidents should be directed towards the causes of accidents and the places where radiation sources are used. Preventing radiation accidents requires effective security measures and a comprehensive approach. The main measures to prevent radiation accidents are:

6.1. Education and Awareness

Education is of great importance in preventing radiation accidents, as in all areas. Many accidents occur due to the inadequacy of qualified and well-trained personnel. Investment in education in areas where radiation is used is more important than investment in equipment. Using equipment without education poses a significant risk (ICRP, 2000).

All personnel working with radiation should receive sufficient training in radiation safety, in addition to being experts in their field. The content of this training should be comprehensive. Topics such as the effects of radiation, methods of protection from radiation, safety procedures, emergency action plans, the use of protective equipment, and the use of personal dosimeters should be covered in these trainings. Furthermore, the continuity of these training courses should be ensured, and the training should be kept up to date according to international guidelines. These trainings should not only be aimed new personnel training but also at updating the knowledge of existing personnel (ICRP, 2000).

6.2. Communication

Especially in environments where there are many different professionals, correct and effective communication plays an important role in eliminating errors. In adequate and effective communication, correct information should be conveyed in a clear manner. In addition to verbal communication, written communication can prevent misunderstandings as much as incorrect transmission of information. In addition, language problems can cause difficulties in communication, so if such a situation exists among employees, special attention should be paid to language-related issues. It should be ensured that the knowledge is transmitted correctly (IAEA STI/PUB/1084, 2000). Specifically in medical radiation accidents, the language difference between the patient and the worker can bring about misunderstandings and errors, so having interpreters in clinics may decrease the number of unexpected events.

6.3. Technical Maintenance and Updates

Radiation devices should be maintained periodically to detect potential risks and prevent failures before they occur.

To increase radiation safety and prevent accidents, existing equipment or devices should be upgraded if necessary or beneficial. Safety upgrades recommended by manufacturers for existing devices should be followed, and mandatory upgrades should be provisioned. Also, old and non-compliant devices must be decommissioned (IAEA SSG-54, 2019).

6.4. Quality Control

Establishing quality control programs and implementing them regularly play a very effective role in preventing radiation accidents. Quality control programs should meet the requirements for quality and safety, and they should be carried out regularly with appropriate equipment. The content of quality control programs ought to be prepared in accordance with the recommendations of international organisations and manufacturers. Controllers recommended by manufacturers have to be regularly carried out at the recommended intervals. Quality control needs to be carried out by trained, competent individuals with appropriate equipment. The controls and calibrations of the equipment used in quality control should also be performed by authorised centers. Proper and effective operation of the quality control mechanism contributes greatly to the prevention of potential errors and accidents (Gonzalez, 1994).

6.5. Documentation

Documentation is crucial for both communication and the transfer of knowledge, as well as for continuity. Educational documents, user manuals, data loaded into treatment devices, acceptance tests, quality control tests, measurements taken, interventions for maintenance and repair, measurements for radiation safety, and information regarding the use, storage, or transportation of radioactive materials should all be documented in detail and stored (IAEA STI/PUB/1084, 2000).

According to the definition of the International Electrotechnical Commission (IEC), detailed user manuals, circuit diagrams as required, calibration instructions, and documents containing instructions for repairing parts that can be repaired by competent personnel have been emphasised as part of the equipment. When these documents are not provided, the equipment is identified as incomplete (IAEA STI/PUB/1084, 2000).

6.6. Incident Learning

Learning from past incidents and preventing their recurrence through reporting is encouraged, and these reports should be analysed, and accident prevention methods should be developed.

In incident learning, personnel should be encouraged to report all incidents, both harmful and harmless, as well as near-misses and unsafe conditions. Encouraging to report all safety-related incidents without fear of punishment must be supported by the management and it should be emphasised that these data are the most important component for creating an effective safety culture in the future. An increase in incident reporting and establishing a strong safety culture reduces the likelihood of accidents. Therefore, documenting these crucial events, learning from them and subsequently developing safety measures will result in both a decrease in the number of accidents and in the severity of these incidents. The option of making anonymous reports should also be provided in open reporting (Zeitman and Palta, 2019).

6.7. Safety Protocols

One of the methods to prevent radiation accidents is the establishment of safety protocols, ensuring the implementation of established programs, regularly reviewing existing protocols, and updating them when necessary.

Particularly, requirements should be imposed to prevent unauthorised people from being present at any stage of the transportation of radioactive sources, such as theft or damage, and the reporting of lost or stolen sources should be mandatory (Gonzalez, 1994).

6.8. Emergency Plans

Accidents or unexpected events can occur despite preventive measures. Therefore, detailed emergency plans for what needs to be done in the event of possible radiation accidents should be prepared. These plans must include instructions for rapid and effective intervention in the event of an accident and necessary steps to minimise the negative effects that may occur during an accident. Emergency plans are to be regularly reviewed and updated when necessary. Relevant personnel should be trained on this matter, and regular drills are needed to ensure that the personnel are capable of implementing these plans (IAEA STI/PUB/1084, 2000).

6.9. Environmental and Physical Security Measures

One of the issues to be considered in preventing radiation accidents is to take necessary environmental and physical security measures against natural disasters or attacks. For this purpose, buildings where radiation sources are located, and nuclear facilities must be constructed to be resistant to natural disasters such as earthquakes, floods, and tsunamis.

Radioactive waste management must be carried out safely to prevent radioactive waste from harming the environment, and these wastes should be stored under appropriate conditions. Also, necessary physical security measures must be taken around buildings where radiation sources and nuclear power plants are located to prevent possible theft, attacks, etc.

6.10. Compliance with National and International Standards

National and international regulations related to radiation safety determine the security practices to be applied in emergencies, and the equipment to be used in accidents and standardise them. Therefore, compliance with these regulations should be ensured, and facilities and applications must be periodically audited for compliance with these standards.

6.11. Monitoring and Reporting

Regular monitoring and control of radiation levels in areas where radiation-emitting devices and radioactive materials are used, and their surroundings are necessary. Through these checks, possible radiation leaks can be detected early, thus preventing possible radiation accidents with early intervention. Also, the calibrations of the devices used in radiation measurements have to be regularly performed, and these devices used in measurements must be traceable.

In the event of any radiation incident or accident, reporting should be done quickly and accurately, and necessary measures must be taken to prevent the recurrence of the accident.

6.12. Technological Advances

The use of new and advanced technologies in preventing radiation accidents provides significant advantages in terms of radiation safety. Automated security systems and remote monitoring technologies can be evaluated in this context.

6.13. Public Awareness

Especially people living near nuclear facilities should be informed and made aware of what to do in case of an accident. In addition, necessary measures should be taken to minimise the effects of radiation on the environment in major radiation accidents.

Conclusion

Prevention of Radiation Accidents is of great importance for both human health and environmental protection. Exposure to radiation can cause serious health problems or genetic damage. It can also lead to environmental pollution and disruption of ecosystems. Therefore, in order to prevent radiation accidents; strict security protocols should be implemented in areas where radiation sources are used, inspections should be regularly carried out periodically, and the continuity of the training of personnel working with radiation should be ensured. These measures taken for radiation protection are of great importance not only for the present but also for the safety of future generations. Prevention of radiation accidents is essential for a liveable world and a sustainable future.

References

Ahearne, J. F. (1980). U. S. Nuclear Regulatory Commission annual report (1979). *Annual report,* No. 5, 1 October 1978-30 September 1979.

ATSDR. (1999). *Toxicological Profile for Ionizing Radiation.* Atlanta (GA). Agency for Toxic Substances and Disease Registry (US)., September 1999. https://www.ncbi.nlm.nih.gov/books/NBK597572/.

Bomanji, J. B., Novruzov, F., & Vinjamuri, S. (2014). Radiation accidents and their management: Emphasis on the role of nuclear medicine professionals. *Nuclear Medicine Communications,* 35(10), 995–1002. https://doi.org/ 10.1097/MNM.0000000000000156.

Burtt, J. J., Akiba, S., Bazyka, D., Coleman, C. N., Hatch, M., & Bernstein, J. L. (2023). Radiation disasters – long term consequences: reflections and summary of a recent symposium. *International Journal of Radiation Biology,* 99(3), 561–568. https://doi.org/10.1080/09553002.2022.2110315.

Cerezo, L. (2011). Radiation accidents and incidents. What do we know about the medical management of acute radiation syndrome? *Reports of Practical Oncology & Radiotherapy,* 16(4), 119–122. https://doi.org/10.1016 /j.rpor.2011.06.002.

Coeytaux, K., Bey, E., Christensen, D., Glassman, E. S., Murdock, B., & Doucet, C. (2015). Reported radiation overexposure accidents worldwide, 1980-2013: A systematic review. *PLoS ONE,* 10(3), 1–26. https://doi.org/10.1371/journal.pone.0118709.

Derreumaux, S., Etard, C., Huet, C., Trompier, F., Clairand, I., Bottollier-depois, J. F., Aubert, B., & Gourmelon, P. (2008). Lessons from recent accidents in radiation therapy in France. *Radiation Protection Dosimetry,* 131(1), 130–135. https://doi.org/10.1093/rpd/ncn235.

Eichholz, G. G. (2004). Accidental Overexposure of Radiotherapy Patients in Bialystok. *Health Physics,* 87(6), 674. https://doi.org/10.1097/00004032-200412000-00026.

Gonzalez, A. (1994). Radiation safety: new international standards. *IAEA Bulletin.* http://scholar.google.com/scholar?hl=en&btnG=Search&q=intitle:Radiation+safety+:+New+international+standards#0%5Cnhttp://www.iaea.org/Publications/Magazines/Bulletin/Bull362/36202040211.pdf.

Hayes, D. F. (1956). A summary of accidnts and incidents involving radiation in atomic energy activities, June 1945 Through December 1955. https://doi.org/10.2172/4343762.

Holmberg, O. (2007). Accident prevention in radiotherapy. *Biomedical Imaging and Intervention Journal,* 3(2). https://doi.org/10.2349/biij.3.2.e27.

Hughes, J. S., & Harvey, M. P. (2006). Radiological Consequences Resulting from Accidents and Incidents Involving the Transport of Radioactive Materials in the UK - 2006 Review (Issue December 2007).

IAEA-INES-2009, 2013. (2013). INES The Internationel Nuclear and Radiological Event Scale User's Manuel. International Atomic Energy Agency Vienna, 2013. https://doi.org/10.2105/9780875532806appv.

IAEA. (2006). Promoting Nuclear Security: What the IAEA is doing. https://www.iaea.org/sites/default/files/nuclsecurity.pdf.

IAEA News, 2011. (2011). Overexposure of workers at irradiation facility - follow up. *IAEA,* 4 (November 2011), 23–26.

IAEA News, 2013. (2013). Exceeding of the maximum regulatory annual effective dose by a practitioner in interventional radiology. 1, 7–8. https://www-news.iaea.org/PrintPreview.aspx?mId=1d022936-bbfb-46e0-b6b5-b05be0dd8cca.

IAEA News, 2024. (2024). Missing dangerous source during transport had been found. May, 8674. https://www-news.iaea.org/ErfView.aspx?mId=fd650b1f-e5df-4c29-8674-64f16b1d610c.

IAEA Non-serial Publications, 2015. (2015). The Fukushima Daiichi Accident Report by the Director General. Director General, 1–222.

IAEA Report INSAG-7, 1992. (1992). The Chernobyl Accident: Updating of INSAG-1. In *Environment International* (Vol. 19, Issue 5). https://doi.org/10.1016/0160-4120(93)90296-t.

IAEA SSG-54. (2019). Accident management programmes for nuclear power plants. Specific Safety Guide SSG-54. *Safety Guide,* 83.

IAEA STI/PUB/1084. (2000). Safety Reports Series 17: Lessons learned from accidental exposures in raiotherapy. *IAEA.* https://www-pub.iaea.org/mtcd/publications/pdf/pub1084_web.pdf.

IAEA STI/PUB/1102. (2000). The radiological accident in Istanbul. International Atomic Energy Agency Publications. https://www-pub.iaea.org/MTCD/Publications/PDF/Pub1102_web.pdf.

IAEA STI/PUB/1644. (2014). Experiences and Lessons Learned Worldwide in the Cleanup and Decommissioning of Nuclear Facilities in the Aftermath of Accidents. *IAEA Nuclear Energy Series,* NW-T-2.7.

IAEA STI/PUB/1776, I. (2018). The Radiological Accident in Chilca. International Atomic Energy Agency Publications, 113. https://doi.org/ISBN:978-92-0-101817-5.

ICRP. (2000). Prevention of accidental exposures to patients undergoing radiation therapy. *Annals of the ICRP,* 30(3), 7–70. https://doi.org/10.1016/S0146-6453(01)00035-5.

International Atomic Energy Agency. (2001). Investigation Of An Accidental Exposue Of Radiotherapy Patients In Panama. Iaea, June. http://www-pub.iaea.org/books/IAEABooks/6391/Investigation-of-an-Accidental-Exposure-of-Radiotherapy-Patients-in-Panama.

Nénot, J. C. (1998). Radiation accidents: Lessons learnt for future radiological protection. *International Journal of Radiation Biology,* 73(4), 435–442. https://doi.org/10.1080/095530098142275.

Saenger, E. L. (1986). Radiation accidents. *Annals of Emergency Medicine,* 15(9), 1061–1066. https://doi.org/10.1016/S0196-0644(86)80130-5.

Sowby, F. D. (1981). Annals of the ICRP. *Annals of the ICRP,* 6(1), 1. https://doi.org/10.1016/0146-6453(81)90127-5.

TAEK Teknik Rapor. (2017). TAEK Teknik Rapor 2017/4: Sakarya ilinde meydana gelen endüstriyel radyografi kazası, müdahale yönetimi ve sonuçları. Türkiye Atom Enerjisi Kurumu. https://kurumsalarsiv.tenmak. gov.tr/bitstreams/7463c638-08db-4d2d-8e29-38afa5022191/download.

Tsegmed, U., Fahim, N., Batcha, A. K., Nakashima, T., Nagata, Y., & Abdel-Wahab, M. (2017). Accidental overexposure related to new radiation therapy technologies. *Journal of Radiation Oncology,* 6(2), 117–132. https://doi.org/10.1007/s13566-017-0309-4.

Turai, I., & Veress, K. (2001). Radiation Accidents : Occurrence , Types , Consequences , Medical Management , and the Lessons to be Learned. *Central European Journal of Occupational and Environmental Medicine,* 7(1), 3–14.

Zeitman A., Palta J. S. (2019). Safety is no Accident: A Framework for Quality Radiation Oncology and Care. 2019. Astro.

Chapter 6

Emergency Aid, Triage Methods, and What to Do in Radiological Accidents

Ayhan Akoz*
and Ahmet Melih Savaş
Department of Emergency Medicine, School of Medicine,
Aydın Adnan Menderes University, Aydın, Türkiye

Abstract

Radiological incidents, not the same as nuclear accidents, are incidental exposures from radiation sources including during transportation, equipment malfunction, or natural disasters, and are known to be recurrent with limited effects. Such incidents can have detrimental health effects such as acute radiation syndrome, cancer, and multi-organ dysfunction, both in the short and long term. Significant exposure was detected in the Goiania accident, which shows how necessary preventive measures are for radiation safety. With the increased occurrence of terror acts, awareness of radiological threats has emerged hence the formulation of response strategies. Radiation emergencies should be prepared for by an integration of planning and preparedness from the emergency management, and public health sections. The mass casualty triage systems such as DIME and START are critical more specifically in mass casualty incidents as they assist in sorting out the patients.

Medical management of radiation exposure entails primary prevention of short-term and long-term health consequences for the individual concerned, as well as management of acute radiation syndrome. It is thus agreed that an emergency management, public health, and

* Corresponding Author's Email: ayhan.akoz@adu.edu.tr.

In: Triage in Radiation Accidents
Editors: Nural Öztürk and Sibel Karaca
ISBN: 979-8-89530-415-0

environmental approach may be deemed as a suitable strategy for both the prevention and management of radiological occurrences. This chapter provides a broad overview of what needs to be done in radiology accidents.

Keywords: emergency preparedness, radiation exposure, radiological incidents, public health response, triage systems

1. Introduction

Radiological accidents encompass numerous circumstances that result in the exposure of human beings to radiation or radioactive matters. These can be transportation, utilisation or loss of radiation sources, having radioactive sources in scrap material, equipment, planning faults, and natural calamities like fires and earthquakes. Such events encompass different aspects of radiation and are not the same as nuclear power plant accidents or nuclear incidents though they have a way of affecting the public through radiation. Radiological events are expected to happen more frequently than nuclear events and have a less severe impact compared to the latter. This implies that radiological accidents happen on a limited scale and the impact is localized, unlike the conventional nuclear disasters which affect vast regions. Maybe they are not frequently encountered, but when they do occur, their effects extend to the environment, public health, and human health (AFAD, 2024).

Accidents related to radiation are fairly described as emergent events that happen due to accidental exposure to ionising radiation when they are over the recommended radiation limits within human health and environmental protection. These events led to a scale of poor health effects including the initial exposure to acute radiation syndrome (ARS), skin necrosis, carcinogenic diseases, and multiple organ dysfunction syndrome. Thus, the consequences of radiation accidents reach not merely health aspects of human beings or even their death rate but also environmental impact with the contamination of ecosystems, disruption of the biodiversities, and ecological integrity. As the consequences, concerning the influence of radiation accidents on human health, the environment, and the whole of society, are broadened, we conclude that the prevention, preparedness, and response to the potential mishap is the inevitable necessity to protect people and minimise negative effects of catastrophes.

A radiation leak occurred in 1987 when a radiation source in an old radiotherapy institute in Goiania, Brazil was opened by accident. Because the source had glow-in-the-dark properties, its contents were disseminated over a wide area, and radiation exposure of 112,000 people had to be evaluated. Thus, after the accident, 249 people were contaminated with radiation, and 4 people died within the firsth month (Anderson and Bokor, 2013). A study aimed at conducting a systemic review of publications and reportage of radiation overexposure cases between 1980 and 2013. There were 634 reported radiation accidents where 2390 people were exposed to radiation and 190 of them died due to radiation exposure. Radiation therapy and fluoroscopy comprised the largest proportion of the reported incidents, 64% in total, which proves the necessity of improving quality assurance in the medical field (Coeytaux et al., 2015).

Further, the emerging issues of terrorism have enhanced privacy concerns especially on radiological disasters, owing to the increased formation of international terrorism groups and the introduction of universally shared knowledge through the internet. This landscape further increases the possibility of deadly attacks using radiological materials; therefore, it becomes important for there to be more effort and work put towards improving the prevention and planning more adequate protective operations in case of these incidents, to the extent of affecting many individuals, the environment, and several sophisticated economies worldwide (Obrador et al., 2022).

The assessment of individuals exposed to radiation and the doses received depends on various factors, which influence the course of radiological incidents in different ways. These factors include such aspects as the geological characteristics of the affected area, the distinction between the environments encompassing large metropolitan areas and those of rural settings, and the features of the construction of buildings and structures defining the degree of protection from radiation penetration. Furthermore, the influence of extraneous conditions comes into the foreground as the key determinant of the effectiveness and intensity of radiation distribution, primarily due to climatic conditions and atmospheric processes that noticeably affect the dynamics of distribution and the kinetics of radioactive substances. In addition, the temporal element is equally vital, as the effectiveness of the first interventions implemented within the critical time frame, including the first hours after the incident, has critical consequences in terms of controlling the extent of radiological contamination and preventing further human and environmental losses. Therefore, the strategies of preparedness and response must be based on a clear and comprehensive understanding of various factors that regulate the behaviors of radiological emergencies, the proactive and increased structural integrity, efficiency of the environmental monitoring systems, and further elaboration of the emergency response plans and actions to be implemented as soon as a radiological emergency takes place. Through the articulation and mitigation of the various external and internal variables influencing the scale and trends of radiological occurrences, all relevant parties can enhance the overall system robustness and ultimately reduce the negative consequences of possible future emergencies (Obrador et al., 2022).

The impact of radiation on human health with due focus on the long-term manifestations has been evidenced by several researchers. Some cohort research studies are useful in explaining the impacts of radiation on carcinogenicity. These findings show that an increased risk of cancer in exposed individuals who receive radiation increases their lifetime risk of developing cancer that is above normal levels. The levels of the effect of radiation on the risk of life-threatening cancers also differ with exposed persons of different age groups where such risks are seen to be higher among exposed children than exposed adults. In this context, several non-cancer effects of radiation exposure such as effects on cardiovascular diseases as well as on some other diseases are also of concern. While high levels of radiation exposure to the body may result in cardiac risk, many experts believe that low-level exposures also have a similar impact, but this has not been confirmed positively. Further studies are recommended to establish the consequences of radiation on health and look for ways of addressing the health issues that arise from radiation (Kamiya et al., 2015).

Clinical symptoms associated with radiation exposure are quite diverse and are outlined in Table 1. These symptoms which include gastrointestinal illnesses to neurological diseases exemplify the impact of radiation as a system disruptive force. Doctors need to be well acquainted with the signs of radiation effects in cases of evaluating and managing the health outcomes of radiation-affected individuals (Reyes et al., 2016).

Radiological incident control and prevention should be a top priority to avoid potential risks and hazardous situations. This involves several key principles, such as the proper and strict use of protective measures, appropriate handling and storage of radiation sources, and the promotion of radiation protection knowledge among the community and radiation workers.

It also involves carefully preparing and executing incident response plans to effectively handle any radiological emergency. An action plan that acts as a guideline for the speedy administration of emergency response, the accomplishment of proper assessment of the victims as well as the coordination of all relevant organisations form the solid foundation to face such an emergency. Having a general strategy for the management and control of over-exposure to radiological-related mishaps is a very reasonable approach in serving the purpose of protecting all the affected individuals.

Table 1. Types of clinical symptoms that may occur due to radiation exposure

CATEGORY	CLİNİCAL MANİFESTATİONS
Gastrointestinal	Abdominal cramps or pain Diarrhea Anorexia
Integumentary	Blistering Erythema Desquamation Onycholysis
Hematologic	Blood loss Granulocyte changes Lymphocyte changes Thrombocyte changes
General Symptoms	Fatigue syndrome Fever Swelling and edema Sensation or itching
Neurologic	Cognitive deficit Headache Neurological deficit
Circulatory	Hypotension
Infection	Increased susceptibility to infection
Dermatologic	Ulcers Necrosis
Excretory	Nausea Vomiting

2. Emergency Aid

Emergency aid in radiation accidents covers a broad spectrum of measures that are vital in an accident to prevent and reduce the consequences. This comprises the activities of identifying and controlling radiation, coordinating evacuations and rescues, decontamination, medical treatment, safety of the involved personnel, psychological impacts, and coordination of the different teams. Taken together, these complex approaches provide responsibilities by stakeholders to fix together and reduce, as considerably as possible, the health and environmental impacts of radiation incidents to secure the lives of affected individuals as well as societies.

The people needed for the first attendance are commonly known as the radiation emergency response team. This team comprises persons who have undergone radiation emergency plans and are part of the drills. The following personnel play important roles in managing radiation accidents: nurses, emergency medicine specialists, trauma surgeons, radiation technicians, security guards, and public relations officers. The specialised staff in advanced care and treatment includes surgeons, radiation technicians, nuclear medicine/radiation therapists, oncology/hematology specialists, and internists. Due to the severe effects of radiation exposure, these specialists are valuable in treating patients exposed to radiation and managing the long-term impacts (Emer, 2017).

The development of emergency response plans concerning radiation disasters must be comprehensive. These plans should include not only medical organisations but also public safety, public health, and emergency management agencies. The following are some of the applications that are important in the management of radiation disasters (Tintinalli et al.,2020).

2.1. Preparation and Organisation of First Responders

2.1.1. Establishing on-Scene Command

Emergency responders must be able to rapidly mobilise and coordinate on-site operations in incidents involving radioactive substances.

2.1.2. Personal Protective Equipment (PPE) and Respiratory Protection

Appropriate PPE and respiratory protective equipment are essential to reduce the risks for the teams.

2.1.3. Casualty Care and Transportation

Transporting and providing care for seriously injured and contaminated patients should be done promptly, considering their medical condition.

2.1.4. Radiation Monitoring and Decontamination

In stable patients, radiation should be measured and decontamination with water or other solutions should be carried out if necessary at the scene

2.2. Communication with Hospitals

2.2.1. Transfer of Incident Information

Emergency responders should convey all information necessary from the scene to the hospitals; this includes incident descriptions, number of victims, nature of injuries, or status of radiation exposure.

2.2.2. Decontamination Information

The status of the patient's decontamination should be communicated before transferring the patient to the hospital.

3. Emergency and Disaster Management in Hospitals

3.1. Hospital Disaster Plan

A hospital must have specific plans for responding to radiological accidents, and for emergency service responders to operate according to these plans. When there is information that a radiation casualty is likely to be brought in, planned activities should be initiated. As many details as possible should be gathered about the casualties, including their medical condition, how they were injured, whether they have been screened for radiation, their current radiological status, the material involved in the contamination, and the expected time of arrival at the hospital. In the case of contamination, the casualty should be considered contaminated until the contamination is ruled out (Emer, 2017).

3.2. Communication with Radiation and Health Physics Experts

Emergency personnel should be aware of how to contact other radiation and health physics professionals in the local area. These specialists are responsible for monitoring radiation dose, inspecting contamination, conducting decontamination, and disposing of contaminated waste.

3.3. Radiation Monitoring and Decontamination

If radiation monitors are not available, decontamination of patients should take place followed by assessment for contamination when suitable monitoring equipment is available.

Each of these elements enhances the safety and efficiency of the community and the healthcare providers in case of a radiation disaster. In this regard, an effective response plan is very important because it guides all the stakeholders on how to respond during a disaster.

4. Triage Methods

In radiation-related emergencies, many casualties enhance the significance of triage processes. Physicians who are likely to treat radiological casualties should know the basic principles of triage, management of radiation exposure, and decontamination to respond to high-level radiation events (Koenig et al., 2005). These processes assist in the coordination of rescue and treatment operations in the best manner possible. The following recommendations can be used in such cases (Tintinalli et al., 2020).

4.1. Implementation of Triage Criteria

4.1.1. Creation of Categories Based on the Status of Patients

Those patients who require medical treatment can be categorised based on the severity of their injuries: mild, postponable, urgent, and difficult to rescue. This categorization should be done regardless of the amount of radiation being administered.

4.1.2. Impact of Radioactive Contamination

Because the effects of radioactive contamination are not immediately fatal, these radiologic evaluations should not impede the early management of emergency-related injuries.

4.2. Radiation and Physical Injuries

4.2.1. Effects of Ionising Radiation

Ionizing radiation, along with other critical illnesses such as physical injuries and burns, contributes to high morbidity and mortality rates. This indicates that these patients require further assessment for radiation exposure in addition to their other injuries.

4.2.2. Complex Emergency Situations

Special situations, including blast injuries with radiation exposure, can be expected to be prioritised because of shortages of resources.

5. Management of Resources

5.1. Developing an Integrated Approach

If there is a lack of resources, hospitals, emergency departments, and other essential establishments should integrate to develop an effective managerial strategy.

5.2. Efficient Use of Resources

Equipment, staff, and treatments should be used to attend to patients based on the severity of the condition, giving priority to those in most critical need of medical attention.

The Radiation Injury Severity Classification (RISC) scale was developed to assess the condition of individuals with acute radiation exposure and classify them into three prognostic categories: Consequently, the definitions of likely survival, possible survival, and unlikely survival are distinguished. The first attempts to estimate the validity of the scale reveal that doctors can distinguish between a large number of situations with a high degree of accuracy, which allows them to state that the scale might be helpful in the identification of consequences after radiation accidents in the framework of emergencies. These results will be validated in larger-scale studies that will be conducted to also enhance the efficiency of the RISC scale (Druzhinina et al., 2010).

DIME classification can be used in mass accidents and characterized by varying levels and can ensure the control of triage and emergency management of radiological and nuclear events. This classification system defines the severity of the situation for the victims and aims to provide appropriate guidance for the effectiveness of the medical response. It is divided into four main categories: "Delayed," "Immediate," "Minimal," and "Expectant." This system is designed to optimise health resources, for example, in cases of mass casualties or radiological attacks. DIME classification is a rapid and efficient way to decide the triage process because it assists the emergency response teams in addressing those individuals who are in more critical condition. This classification which is used during treatment in mass radiation accidents is helpful to get maximum results because opens the way to responders. The DIME classification is presented in Table 2 (Ayan and Dönmez, 2018).

In the case of radiological incidents, it is crucial to have an effective means by which to address a large population of patients in a short amount of time, and the START triage algorithm is one such method. It focuses on people who need immediate treatment to save lives or limbs; it makes a brief evaluation of mobility, respiratory state, perfusion, and mental state. Using the color-code system, patients who require immediate attention are moved to the red team, while those who do not require urgent attention but must be seen quickly are tagged yellow. Those with minor injuries that might not require surgery are labeled with a green wrist tag. This system ensures that the utilization of scarce medical resources is optimized and used in a way that the most number of lives are saved during the crisis which is a standard formula used because of the magnitude of the disaster (Bazyar, Farrokhi, and Khankeh, 2019).

The Jump START triage is an important aspect of the pediatric management plan in mass casualty incidents since children have different physiology from adults. It adapts an adult-based START triage system to the pediatric one regarding the different physiological parameters and developmental levels of children. For example, Jump START advises checking for a pulse and administering five rescue breaths to an apneic child who has a pulse, which is consistent with a higher probability of respiratory emergencies in children. The system focuses on quick evaluation and sorting as the first responders must prioritize the distribution of scarce resources and help as many children as possible (Jones et al., 2014).

Table 2. DIME Classification

Category	Condition Description	İntervention Priority
DELAYED	Stable but requires monitoring, non-life-threatening injuries (e.g., non-severecuts, fractures)	Non-urgent, canbedeferred
IMMEDIATE	Individualsrequiring urgent intervention for vital functions (e.g., severe respiratory distress, unconsciousness)	Immediate intervention required
MINIMAL	Individualswhocan move independently and have minor injuries	Lowpriority for intervention
EXPECTANT	Those unlikely to survive or with mortal injuries	Lowest priority, comfortcare-focused

The SALT triage system (Sort, Assess, Lifesaving Interventions, Treatment/Transport) is a complete system for the management of mass casualty incidents (MCI). SALT can be used in general MCI and for more specific events such as the ones involving radiological exposure and is a rapid and effective way of categorising victims starting with the division of people into walking and non-walking. Emergency procedures are applied, if necessary, but do not delay the triage process. This system highlights the importance of quick decision-making on who should be treated or transported first to benefit as many victims as possible (Bazyar, Farrokhi, and Khankeh, 2019).

In a study, an automated micronucleus scoring method was designed for rapid population sorting in large-scale radiation incidents. With the help of the specially developed software module, it was possible to evaluate 60 blood samples per day, and the radiation dose of 1 Gy could be assessed with a vagueness of 0.2Gy. Further comparison of the results obtained in different laboratories in Belgium and Germany in the context of a multicenter test also proved the effectiveness of the method and its applicability for rapid differentiation in several centers. This implies that automated micronucleus scoring could be helpful as a fast way of sorting out large populations after a radiation disaster (Willems et al., 2010).

The measurement of dose has taken a central position in the assessment of exposure to radiation during accidental events and the formulation of the strategies to be used in treating the victims respective of the kind of radiation incurred. In this connection, it is possible to identify several biodosimetric methods, which are conceived as primary means of assessing the degree and level of irradiation. They include methods like the dicentric assay, micronucleus assay, gamma-H2AX foci assay, and optically stimulated luminescence each carrying its own measure of strength and disadvantage. The dicentric assay gives the best results with a very high level of precision in most cases and the micronucleus assay can give faster results depending on the situation. The gamma-H2AX foci assay has the potential of viewing the biological impacts present as DNA damage; however, this technique needs to be performed as soon as possible after sample collection. Optically stimulated luminescence is highly proficient over large dose ranges and is specifically preferred when determining high-dose impacts. Table 3 below gives the comparative advantage and drawback of the approach listed above and as discussed above, it highlights the suitability of the method in the scenarios associated with the radiation accident (Ainsbury et al., 2014).

These targeted strategies as outlined above can be very helpful in reducing the number of deaths, particularly in fatal incidences caused by radiation disasters. Comprehensive assessment of external and internal physical wounds and other health issues that may grow because of radiation exposure is vital in the triage process. Another point that should be underlined is that not only the members of emergency service teams who will be working in such conditions have to understand such conditions, but they have to prepare themselves for working in such conditions as well. This implies the need for quick and synchronised response ability which are part of crisis mitigation measures. Therefore, to ensure that these teams are well equipped, trained, and prepared to respond to these incidents more efficiently and in a coordinated manner, hence improving on the general handling of these issues, more lives can be saved, and there will be reduced long-term health repercussions.

Table 3. Comparative analysis of biodosimetric methods used in radiological accidents

Biodosimetry Method	Benefits	Challenges
Dicentric Assay	Offers high accuracy and reliability.	Processing can be lengthy and labor-intensive.
Micronucleus Assay	Can produce fast results.	May be less sensitive at low dose levels.
Gamma-H2AX Foci Assay	Can directly show DNA damage.	Requires rapid processing after sample collection.
Optically Stimulated Luminescence	Effective over a wide dose range.	Requires complex equipment and expertise.

6. What to Do

Radiation injuries do not normally pose an immediate threat of death, which gives healthcare professionals some time to determine whether the patient has been exposed to radiation and whether it is external or internal contamination. Information provided from biodosimetry, initial clinical presentations, and relevant laboratory abnormalities observed during the first 24 to 48 hours of emergency patient evaluation informs early interventions (Berger et al., 2006; REAC/TS, 2007).

Healthcare staff have a small potential for exposure to radioactive material if a patient is contaminated. Nevertheless, the purpose of decontamination is to bring the level of radiation as low as reasonably achievable so that both the patient and the healthcare personnel are exposed to the least amount of radiation. This is done by shortening the exposure time, minimising the amount of the radiating materials in the particular environment, distancing the source, and utilising shielding (Berger et al., 2006; REAC/TS, 2017).

Patients who are exposed to ionising radiation require special treatment in managing ARS. In managing ARS, the goal is to address prodromal phase symptoms. See Table 4 for recommendations. (Tintinalli et al., 2020).

Table 4. The management of acute radiation syndrome

Analgesia should be provided.
If the exposure dose is above 5 to 6 Gray (Gy), NSAIDs should be avoided because of the risk of serious gastrointestinal bleeding. (Christensen, Iddins, and Sugarman 2014)
Antiemetics such as ondansetron are recommended for nausea and vomiting. (Ricks, Berger, and O'Hara 2002)
For diarrhea management, loperamide can be used if necessary.
Fluoroquinolones may be used for the management of GI syndrome 2 to 4 days after acute exposure.

In the process of treatment, it is relevant to conduct a detailed anamnesis and physical examination. Specific attention should be paid to the early manifestations of acute radiation syndrome, including the time of vomiting and diarrhea, fever, hypotension, tachycardia, and tachypnea (Tintinalli et al., 2020). All laboratory tests should be done quickly to provide biological dosimetry. Other laboratory data include the rate of decline in lymphocyte count and the nadir, as well as cytogenetic study for chromosome aberrations (Ricks, Berger, and O'Hara, 2002).

ARS management entails maintenance and support of the hematologic system, control of infection, and possibly management/treatment of multiple organ dysfunction syndrome. In some cases, antibiotics, antifungals, and antivirals may be required for the prevention of infections. Also, patients should be carefully monitored and managed for severe infections and metabolic complications that may occur in the course of high radiation doses. Supportive interventions include early surgery and wound care involved in this process (Tintinalli et al., 2020; Ricks, Berger, and O'Hara, 2002).

Some of the life-threatening complications of high-dose radiation exposure include ARS leading to multi-organ involvement and multi-organ failure. They are associated with systemic inflammatory response syndrome (SIRS). As cells are damaged by radiation, they become ineffective and lead to deterioration in organ functions. The management approaches are typically oriented toward the restoration of homeostasis and the recovery of the functions of the referred organs. However, as radiation accidents are not that frequent, there is still not much information on the management of such conditions, as the majority of the information is based only on several controlled clinical trials. These limitations are mostly overcome by information obtained from experimental studies using animal models. Many of the current practices of radiation injury management have now emerged as the outcome of these experimental studies. Thus, the new therapies and the existing ones, international cooperation, and interdisciplinary teamwork are of great significance. The new strategies to manage local radiation injuries, such as mesenchymal stem cells, have their advantages. These methods could be potentially helpful in avoiding late radiation toxicity due to the stimulation of tissue regeneration and couching inflammation. In conclusion, more studies and investigations in this area are needed to grasp higher knowledge and better management of multiple organ failure provoked by radiation-related accidents (Dörr and Meineke, 2011).

The management of local radiation injury majorly entails careful wound care and attention to pain control as well as infections. Cutaneous radiation injury is not a thermal burn and the injury can progress over time. The first intervention is to reduce the inflammation in the skin that has been affected by radiation. Topical steroids, oral vitamins, and pentoxifylline are some of the agents that can be administered to increase blood flow (REAC/TS, 2017; Dainiak et al., 2011). Other treatments include silver-containing creams like sulfadiazine and hyperbaric oxygen therapy may also be used. The use of systemic steroids is not recommended (Dainiak et al., 2011).

There are severe consequences of radiation exposure which may cause deep tissue injury and loss of several organs, so it is crucial to find ways of intervention in such cases. Surgery is the primary modality of treatment for severe radiation injury and common operations include excision of necrotic tissues, skin grafting, and vascularised flaps. In more recent years, new treatment methods employing the patient's own stem cells have also been of interest. Notably, adipose tissue, which is procured through less invasive methods and contains a large number of mesenchymal stem cells, can be considered a promising source in this field. Clinical practices have confirmed the effectiveness of such methods, particularly in patients with severe health issues. Furthermore, research on the molecular mechanisms of the injury induced by radiation enables the creation of new therapeutic approaches regarding single nucleotide polymorphisms and radiation fibrosis. These genetic and molecular strategies provide the basis for the establishment of individual treatments in cases of radiation exposure and accidents and create new opportunities for the prevention of the late effects of radiation exposure (Akita, 2014).

Internal contamination does not always manifest symptoms, however, if there are constantly high readings of radiation, symptoms should be heeded. For detection of the possible radionuclide, it is advised to perform a 24-hour urine test. Cathartics, activated charcoal, gastric lavage, and decorporation agents selectively for certain radionuclides can be given. Treatment is done according to dose estimates that were obtained through radiochemical analysis (Tintinalli et al., 2020).

Conclusion

Radiological accident management is a multifaceted issue that cannot be solved by one discipline only, and it includes first intervention and triage, disease prevention and control, and aftercare. Each of these components is crucial in the event of such an accident. These core interventions include providing easy access to initial support for those in need or prioritising help based on injury severity. Additionally, other measures in the field of public health are planned to prevent the spread of radiation, and constant evaluations are conducted to determine its impact on the environment and people. The rapid response to such incidents is also possible due to adequate preparedness; having a proper plan in place significantly reduces response time. Taking prompt action is extremely important to prevent the situation from deteriorating further. Other skills like communication are just as important. Effective, timely, and accurate communication during calamities helps reduce public stress and confusion, guiding affected individuals on how to protect themselves from contamination and where to seek assistance. Significant emphasis should be placed on all these factors to minimize the impact of a radiological disaster on both public health and the environment.

References

AFAD. (2024). *"Radyolojik Tehdit ve Tehlikeler."* n.d. Accessed January 26, 2024. https://www.afad.gov.tr/kbrn/radyolojik-tehditler.

Ainsbury, E. A., Al-Hafidh, J., Bajinskis, A., Barnard, S., Barquinero, J. F., Beinke, C., ... & Wojcik, A. (2014). Inter-and intra-laboratory comparison of a multibiodosimetric approach to triage in a simulated, large scale radiation emergency. *International journal of radiation biology*, *90*(2), 193-202.

Akita, S. (2014). Treatment of radiation injury. *Advances in wound care,* 3(1), 1-11.

Anderson, P. D., & Bokor, G. (2013). Nuclear and radiological terrorism: continuing education article. *Journal of pharmacy practice,* 26(3), 171-182.

Ayan, A., & Dönmez, S. (2018). Radyolojik Nükleer Kaza ve Terör Olaylarında Tıbbi Yönetim. *Ankara Eğt. Arş. Hast. Dergisi,* 51(2), 154-162.

Bazyar, J., Farrokhi, M., & Khankeh, H. (2019). Triage systems in mass casualty incidents and disasters: a review study with a worldwide approach. *Open access Macedonian journal of medical sciences,* 7(3), 482.

Berger, M. E., Christensen, D. M., Lowry, P. C., Jones, O. W., & Wiley, A. L. (2006). Medical management of radiation injuries: current approaches. *Occupational Medicine,* 56(3), 162-172.

Christensen, D. M., Iddins, C. J., & Sugarman, S. L. (2014). Ionizing radiation injuries and illnesses. *Emergency Medicine Clinics,* 32(1), 245-265.

Coeytaux, K., Bey, E., Christensen, D., Glassman, E. S., Murdock, B., & Doucet, C. (2015). Reported radiation overexposure accidents worldwide, 1980-2013: a systematic review. *PloS one*, *10*(3), e0118709.

Dainiak, N., Gent, R. N., Carr, Z., Schneider, R., Bader, J., Buglova, E., ... & Meineke, V. (2011). Literature review and global consensus on management of acute radiation syndrome affecting nonhematopoietic organ systems. *Disaster medicine and public health preparedness*, *5*(3), 183-201.

Dörr, H., & Meineke, V. (2011). Acute radiation syndrome caused by accidental radiation exposure-therapeutic principles. *BMC medicine*, *9*, 1-6.

Druzhinina, M., Kuniak, M., Azizova, T. V., Zhang, A., Day, R., & Wald, N. (2010). Prediction of the Severity of Acute Radiation Syndrome Severity. *Prehospital and Disaster Medicine*, *25*(S2), s94-s95.

Emer, M. O. (2017, November). Hospital Preparation in the Radiological/Nuclear Urgents and Organization of the Emergency Response Team/Radyolojik/Nukleer Acillerde

Hastane Hazirligi ve Acil Durum Mudahale Ekibi Organizasyonu. In *Nuclear Medicine Seminars* (Vol. 3, No. 3, pp. 225-231). Galenos Yayinevi Tic. Ltd..

Jones, N., White, M. L., Tofil, N., Pickens, M., Youngblood, A., Zinkan, L., & Baker, M. D. (2014). Randomized trial comparing two mass casualty triage systems (JumpSTART versus SALT) in a pediatric simulated mass casualty event. *Prehospital emergency care*, *18*(3), 417-423.

Kamiya, K., Ozasa, K., Akiba, S., Niwa, O., Kodama, K., Takamura, N., ... & Wakeford, R. (2015). Long-term effects of radiation exposure on health. *The lancet,* 386(9992), 469-478.

Koenig, K. L., Goans, R. E., Hatchett, R. J., Mettler Jr, F. A., Schumacher, T. A., Noji, E. K., & Jarrett, D. G. (2005). Medical treatment of radiological casualties: current concepts. *Annals of emergency medicine*, *45*(6), 643-652.

Obrador, E., Salvador-Palmer, R., Villaescusa, J. I., Gallego, E., Pellicer, B., Estrela, J. M., & Montoro, A. (2022). Nuclear and radiological emergencies: biological effects, countermeasures and biodosimetry. *Antioxidants,* 11(6), 1098.

REAC/TS. (2017). The Medical Aspects of Radiation İncidents. *REAC/TS*. 4. Edition. Oak Ringe.https://orise.orau.gov/resources/reacts/documents/medical-aspects-of radiation-incidents.pdf.

Reyes, E. H., Baciu, F., Benderitter, M., Lataillade, J. J., Bey, E., Trompier, F., & Tamarat, R. (2016). Medical response to radiological accidents in Latin America and international assistance. *Radiation Research*, *185*(4), 359-365.

Ricks, Robert C., Mary E. Berger, and Frederick Michael O'Hara, eds. (2002). *The Medical Basis for Radiation-Accident Preparedness: The Clinical Care of Victims: Proceedings of the Fourth International REAC/TS Conference on the Medical Basis for Radiation-Accident Preparedness, Held March 6-8, 2001, in Orlando, Florida*. Boca Raton: Parthenon Pub. Group.

Tintinalli, J. E., Stapczynski, J. S., Ma, O. J., Yealy, D. M., Meckler, G. D., & Cline, D. M. (2016). *Tintinalli's Emergency Medicine: A Comprehensive Study Guide,* 8e (Vol. 18). McGraw Hill Education.

Willems, P., August, L., Slabbert, J., Romm, H., Oestreicher, U., Thierens, H., & Vral, A. (2010). Automated micronucleus (MN) scoring for population triage in case of large scale radiation events. *International journal of radiation biology,* 86(1), 2-11.

Chapter 7

Emergency Aid, Triage Methods and What to Do in Nuclear Accidents

Ali Duman*
and Ahmet Melih Savaş
Department of Emergency Medicine, School of Medicine,
Aydın Adnan Menderes University, Aydın, Türkiye

Abstract

Nuclear disasters may cause severe difficulties due to immediate radiation danger and irreversible impact on nature and health. The events in Chernobyl and Fukushima serve as a reminder of the critical importance of emergency medical response and hospital competency for disasters. In emergencies, a prompt and efficient response will lead to a strong public health system that will help communities withstand radiation exposure and recover quickly. Strategies for community health care surveillance and methods for assessing triage conditions should be optimised through proper emergency preparedness in the event of nuclear disasters. Effective communication and collaboration among various medical specialties, including radiation oncologists, nuclear medicine specialists, emergency medicine specialists, and other relevant physicians, is vital for ensuring prompt and coordinated treatment and response in the event of a nuclear crisis. The comprehensive approach described helps minimise the impact of nuclear accidents by improving the speed and effectiveness of emergency medical response. Developing emergency response programs is crucial for creating a more stable society following a

* Corresponding Author's Email: ali.duman@adu.edu.tr.

In: Triage in Radiation Accidents
Editors: Nural Öztürk and Sibel Karaca
ISBN: 979-8-89530-415-0

nuclear disaster. This section of the book offers a detailed report on how to handle nuclear accidents.

Keywords: acute radiation syndrome (ARS), emergency medical interventions, medical triage methods, nuclear accidents, radiation exposure

1. Introduction

The term "nuclear hazard" encompasses the potential dangers associated with the release of energy from the fusion of atoms or fission of atomic nuclei, leading to destructive effects. These hazards involve the presence of violent pressure waves, lethal radiation levels, and extremely high release temperatures. The lethal radiation impact of a nuclear explosion is very diverse. The impact of a nuclear explosion is far-reaching, generating intense light, heat, pressure waves, and radioactive contamination that can spread into the air, water, and soil, causing widespread destruction in the surrounding area (AFAD, 2024).

The Chernobyl nuclear power plant catastrophe is considered the biggest in the history of technological accidents. Estimates suggest that there were between 5,000 and 75,000 cases of radiation-induced cancer, and approximately 130,000 people were medically evacuated (Byrne and Hoffman, 1987). On the other hand, the Fukushima Daiichi nuclear power plant disaster is considered to be a cataclysm, with the death toll reached 2,313 disaster-related deaths among evacuees from the Fukushima area (World Nuclear Association, 2024).

The lessons from the Fukushima Daiichi nuclear accidents are crucial for developing effective emergency medicine approaches in nuclear accidents. In particular, it is important to have cooperation between national and local authorities, rapid and effective evacuation strategies, appropriate communication during the crisis, and preparation plans and exercises. These factors increase the effectiveness of emergency medical interventions in nuclear accidents (Plants et al., 2014).

The event of the nuclear explosion of the Fukushima Daiichi Nuclear Power Plant shows how important emergency medical facilities and triage techniques are in risky scenarios such as nuclear catastrophes. The Japanese Society of Acute Medicine (JAAM) created an emergency task force that provides medical support to on-site personnel exposed to the risks of radiation and other secondary mishaps. In total, 261 patients were treated, and none of them developed indicators of acute radiation syndrome, but eight were contaminated externally. JAAM was able to apply the methods of triage and organise the delivery of emergency medical services and patient management. This illustrates the need for care in triage and medical help in future nuclear catastrophes or disasters (Morimura et al., 2013).

Radiation accidents are less frequent than nuclear accidents; however, the aftermath of nuclear plant accidents is more disastrous (Ovalı, 2021). In addition to the loss of lives and immediate destruction, the use of nuclear weapons creates long-term environmental and health issues. Radioactive contamination can affect humans, animals and plants, and its consequences can take longer to develop compared to other kinds of contamination. Incidents like this can pose great danger to health and the environment, therefore, a quick and decisive response is necessary to prevent severe outcomes.

Studies have shown that people living in areas affected by nuclear incidents face serious social challenges. These challenges include joblessness, separation, lack of social support, conflict, concerns about radioactivity, and other issues that greatly impact the overall quality of life and psychological well-being of the people in the region. It is crucial to raise awareness about radiological regulations and to encourage community involvement to address these challenges and move towards recovery. This underscores the importance of community-oriented approaches and the active involvement of locals in conducting social and personal investigations following nuclear catastrophes (Maître et al., 2021).

2. Emergency Aid

After a nuclear radiation disaster, hospitals should be all ready to do several procedures. Hospitals are recommended to carry out environmental risk assessments on a regular basis and make sure that there are adequate staff and equipment that meet the identified risks. The need to collaborate with the fire department, security agencies, and other relevant ministries, as well as establish a certain level of rapport with these units, is crucial (Emer, 2017).

When it comes to the impact of radiation on public health, the effects are similar but can lead to health issues that are not necessarily directly caused by radiation exposure. However, it is difficult to conduct studies on successful planning and response techniques due to the infrequency of disasters. Therefore, it's important to identify lessons learned and implement measures to protect emergency services teams and the public. In addition to providing healthcare operations such as medical and non-medical care and psychological support, healthcare workers also need training in communication and nuclear incident response strategies (Ohtsuru et al., 2015).

While medical emergencies being caused by nuclear accidents immediate effect will be of acute health conditions because of radiation exposure, the worse side effect will, nevertheless, be the result of displacements and forced evacuations of the already vulnerable population, like very old people and patients who are hospitalised and in need of regular care. The Fukushima Daiichi nuclear power plant accident can be taken as a typical case that shows the difficulties of emergency healthcare services, and the requirement of healthcare emergency response strategies (Arifumi Hasegawa et al., 2015).

The SHAMISEN project in 2021 has made significant recommendations that will help enrich the health monitoring strategies and response to nuclear accidents. These recommendations are the quick and correct assessment of radiation dose in an emergency, implementing long-term health monitoring programs, and dealing with psychological problems of the individuals. As such, the staff will be trained to handle radiation information and psychological support, develop empathetic communication, and incorporate the community in health surveillance processes. Besides, SHAMISEN suggests that the practice of community-based methodology should be introduced and community leaders, health experts, and health officials must be involved in this procedure. Executing the setups would result in building comprehensive and effective approaches that will meet the needs of survivors better and the community's recovery process will be quicker after the nuclear accidents (Ohba et al., 2021).

It is of primary importance that the emergency service measures should be well prepared in the case of a nuclear accident because the aftermath of such event can become a disaster. The recommendations below concern these cases (Tintinalli et al., 2020).

2.1. Commencement of Hospital Disaster Plan

2.1.1. The Activation of the Radiation Experts

Preparing the hospital's radiation safety officers, nuclear medicine, radiation oncology department and the personnel in related areas of work for emergencies.

2.2.2. Request for Dosimeter

The establishment of dosimeters, radiation monitoring, and research instruments for the personnel.

2.2. The Planning of Preparation of the Emergency Room

2.2.1. Establishing a Temporary Triage Area

The creation of an emergency triage zone for the quick team assessment of the victims and, consequently, their direction to the branch of the hospital, issued in the hospital action plan.

2.2.2. Identification of Contaminated and Clean Areas

The development of a clear and effective separation zone between the contaminated and clean areas.

2.2.3. Prevention of Contamination

It is suggested that people do not carry their external apparel with them out from the contaminated area, and do a radiation check before getting out of such place.

2.2.4. Protection of the Floor

Covering the floors with plastic or paper and securing them with heavy tape.

2.2.5. The Protection of People at Risk

It is recommended that pregnant women and individuals who should not remain in the area evacuate.

2.2.5.1. Protection of Staff

2.2.5.1.1. Use of Waterproof Protective Clothing

The staff must wear a waterproof apron, headgear, and shoe cover.

2.2.5.1.2. Double Glove Usage

It is recommended to tape the inner glove and replace the outer pair after contact with contaminants and between patients.

2.2.5.1.3. Face Mask Usage

For optimal protection, N95 masks are recommended. In the event of unavailability, surgical masks are a suitable alternative.

2.2.5.1.4. Measurement of Radiation

It is advised that constant checks are to be made on the hands and clothes, with a radiation meter.

2.2.5.1.5. Dosimeter Usage

Dosimeters must be available and must be worn on the collar under protective clothing.

Nuclear medicine staff are the appropriate team to assist emergency medicine teams in the event of a radiation incidente. Additionally, the drills simulate the readiness of organisations for emergencies and facilitate collaboration with external agencies (Schleipman, Gerbaudo, and Castronovo, 2004).

The types of nuclear accidents vary, ranging from radioactive contamination to nuclear explosions, and therefore special preparations must be made by medical departments for such emergencies. Nuclear medicine departments should consider accidents that might take place in the hospital service area. This is particularly the case for hospitals in the vicinity of significant nuclear power plants or research facilities. Departments of nuclear medicine are crucial to assess the degree of radioactive contamination and control the decontamination process. In this regard, standard procedures of possible accidents should be developed in collaboration with emergency department as well as the concerned nuclear facilities. These activities should be complemented by frequent training sessions and should be included in the hospital emergency plans. Consequently, quick and efficient triage and treatment can be given in nuclear accidents (Paans and De Jong, 2017).

The management of nuclear disasters requires collaboration among doctors with various specialties. Effective communication among professionals such as radiation oncologists, hematologists, medical practitioners, psychologists, and surgeons is crucial for the successful treatment of victims, addressing both their physical and mental health needs. The multidisciplinary approach enables swift and effective response to emergencies and improves long-term health outcomes. Additionally, a multidisciplinary team can provide a comprehensive approach to addressing the complex health issues resulting from radiation accidents and develop individualized treatment plans for each victim. Therefore, fostering cooperation is a top priority in post-nuclear settings, as it enhances the quality of life for victims and the effectiveness of treatments. (Bomanji, Novruzov, and Vinjamuri, 2014).

The evacuation of hospitals and care facilities in case of a nuclear accident can be the toughest task, especially if there is no proper planning. To illustrate, with the Fukushima incident ; more than 60 patients and elderly residents of care facilities died during or after evacuation because of insufficient planning. Hence, it is crucial to have complete and multi-faceted evacuation plans for hospitals and care facilities to review them often and conduct drills. Besides, it is essential to have resources like generators, mobile communications equipment, and portable medical supplies in advance to guarantee emergency medical services continuity. This kind of readiness increases the safety of both patients and medical workers and at the same time, it provides a good health response in the messy conditions of a nuclear accident (Ohba et al., 2021).

These measures aim to improve the safety of both patients and healthcare workers in case of a radiation disaster. The safeguarding of people from the possible dangers of radiation is a basic part of disaster response.

3. Triage Methods

The systems of triage that are used in the wake of nuclear disasters consist of emergency medical assistance, quick assessment, contamination control, and detailed triage. The triage technique is a practical way to categorise the injured people in nuclear and radiation accidents and their rescue, thus reducing the risks. A lot of triage systems are used for nuclear radiation-injured people. Because there is no single, universally or nationally instituted standardized system, it is essential to have a triage and decontamination methodology that is easy to understand and can be adjusted to the situation, such as the availability of medical resources, the number of casualties, and the severity of injuries. These methods focus on the cure of individuals, considering radiation exposure and any related physical injuries. The procedure implies the following steps (Karaçalıoğlu, 2017).

3.1. First Aid and Emergency Medical Care

The first step in the decontamination process is to scan victims using radiation detectors to determine the level of radiation exposure. Those identified as contaminated must then be thoroughly washed. First responders are trained to use protective equipment to prevent exposure through the respiratory tract, skin, and eyes.

3.2. Personal Protective Equipment and Safety

Healthcare personnel involved in nuclear and radiological situations must be well-versed in radiation protection protocols and have all necessary protective equipment. This includes implementing measures to prevent radiation-related harm.

3.3. Decontamination and Second Scanning

Patients transported to the first aid center undergo a secondary assessment based on their physical condition and mobility. At this stage, any traumatic injuries and urgent surgical requirements are also addressed. Those without trauma are evaluated for symptoms related to radiation exposure.

4. Clinical Evaluation and Follow-Up

The survivors' health is carefully assessed through chromosome analysis and full blood counts. These tests reveal the extent of radiation exposure and help determine the necessary treatment and follow-up procedures.

In the event of a nuclear disaster, it is crucial to have biodosimetry tools that can quickly determine radiation doses at the individual level to enable effective medical triage. Initial triage is based on the patient's location, symptoms, and physical examination, while more detailed assessments are conducted to accurately determine radiation levels. It is unreasonable to expect one biodosimetry test to address all the needs of multi-system radiation injuries due to their complex nature. To ensure future technologies are tailored to patient needs, it is important to consider factors such as sample type, dose detection limit, analysis time, ease of use, and contribution to the treatment process (Sullivan et al., 2013).

Triage tags are highly valuable in emergency management for large-scale disasters. These tags allow first responders to quickly assess and separate casualties based on whether they require immediate intervention. It is important to choose water-resistant triage tags that allow for easy labeling of information. Advanced systems like Medical Emergency Triage Tags (METTAG) are designed for use in major accidents or terrorist attacks, helping to effectively manage resources in such scenarios. Contemporary triage tags such as CB-100 are made of water-proof, high-density materials, enabling quick note-taking and efficient monitoring of patient conditions. Each part of the tag has its own barcode number, which is used to track the triage process for patients. The use of these innovative triage tags can enhance the speed and accuracy of emergency response, ensuring timely and accurate treatment for patients (Ramesh and Kumar, 2010).

Regarding the application of triage methods in the case of a nuclear accident, the effective application of the said methods is paramount. The START and MSTART systems are appropriate when rapid evaluation and immediate action are needed. The START system is designed to quickly assess the casualty using the elements of ability to walk, respiratory status, or compliance with commands. This helps in the enhancement of intervention rate and quick action so as to deal with critical cases in nuclear accidents. The changes in MSTART enable better triage in the dark and cold conditions, especially by excluding the capillary refilling observation. The second part of SALT classification takes less time to perform as it does a rapid classification with simple voice commands and concentrates on interventions that are lifesaving, especially if a patient is in need of urgent assistance. These systems ensure the effective use of healthcare in complex situations arising from nuclear accidents. In the case of nuclear incidents, it is proposed that these systems can be used to help determine the degree of contamination and identify radioactively contaminated individuals. Table 1. below summarises the main features of the START, MSTART, and SALT triage systems and their applications in nuclear accidents (Bazyar et al., 2019).

The proper evaluation of radiation doses in survivors and the triage of affected individuals after a nuclear accident present significant medical challenges. Sequential complete blood counts (CBCs) are the most effective method for assessing radiation exposure. Nevertheless, an approach like this does not comply with the required scale at once. A study suggests the Exposure and Symptom Triage (EAST) tool, which is a combination of exposure estimates and clinical evaluations. This instrument can be useful to triage and deliver first aid to the survivors in the damaged areas (Hick et al., 2018).

Table 1. The main features of the START, MSTART and SALT triage systems

Triage System	Key Features	Application in Nuclear Accidents
START	Quickly assesses casualties based on ability to walk, respiratory status, or command compliance.	Enhances the rate of intervention and quick action to manage critical cases effectively. Suitable for rapidevaluation and immediate action in nuclear accidents.
MSTART	Adapted for dark and cold conditions, excludes capillary refill observation.	Enables better triage under challenging conditions such as darkness and cold. Designed to improve the accuracy and speed of triage in adverse environments.
SALT	Uses simple voice commands for rapid classification, and focuses on life-saving interventions.	Takes less time to perform and emphasises rapid classification with immediate life-saving interventions. Efficient in identifying and prioritising patients in urgent need of assistance.

The Radiation Triage, Treatment, and Transport (RTR) model is the functional medical response model for nuclear and radiological mass casualties. It entails radiation-specific triage, treatment, and transport sites that are selected on the characteristics of the event and the environment. The RTR system is designed to provide medical care after a nuclear incident. It defines three types of RTR sites based on different radiation levels and environmental conditions. RTR1 includes sites with high radiation and mostly inpatient patients; RTR2 involves sites with less radiation and mostly ambulatory patients; and RTR3 involves sites with minimal radiation risk and the best transport facilities. Continuous radiation monitoring is essential for the effective functioning of this model, ensuring the safety of emergency teams. This helps in the prompt and effective delivery of emergency care and triage, as well as minimising the risk of harmful ionising radiation exposure to medical personnel and casualties. The Table 2 below summarises the RTR model (Hrdina et al., 2009).

5. What to Do

Measures are necessary to ensure the health and safety of nuclear accident victims and communities. This can be done through the use of successful disaster preparedness and management measures aimed at minimising the impact of such occurrences. This approach involves several measures such as chaos control, the development of good evacuation plans, and the organisation of emergency responses. The following overview presents the principal preparedness and response methods to be applied in nuclear radiation exposure situations (Walls et al., 2023).

5.1. Emergency Preparedness & Communication

5.1.1. Community Disaster Plan

A plan must be developed by identified and authorised personnel and should include the following elements: evacuation, treatment, and public information.

Table 2. The Radiation Triage, Treatment, and Transport (RTR) Model

RTR Site Type	Radiation Level	Characteristics
RTR1	High Radiation	Require inpatient care
RTR2	Lower Radiation	Mostly ambulatory
RTR3	Minimal Radiation	Best transfer facilities

5.1.2. Dissemination of Information

Provision of correct and timely information to the general public ensures that chaos does not reign. Therefore, the media should receive the information through a public relations officer.

5.2. Radiation Control Staff

Personnel in charge of radiation control at the hospital or the incident site should ensure that all patients and medical personnel are monitored using a radiation counter (e.g., Geiger counter). This is to determine whether contamination has occurred and to control cleaning and diversion if necessary

5.3. External Contamination Intervention

5.3.1. Emergency Medical Interventions

The non-immediate life-threatening nature of radiation contamination necessitates the implementation of first aid measures.

5.3.1.1. Decontamination

If a patient is contaminated, appropriate decontamination procedures are critical. These measures include general precautions such as wearing rubber gloves, shoe covers, and medical masks if required.

External decontamination is a process which is targeted at the elimination of the radioactive materials from the skin and clothing of individuals. This process commences by removal of the hazardous clothing and washing of the surfaces of contaminated individuals with soap and water. The decontamination procedure removes radioactive particles from the skin and open wounds, minimising further radiation exposure to the affected person and those around them. The effectiveness of the decontamination procedure relies on the precision and swiftness of the techniques used (Flynn and Goans, 2006).

5.3.1.2. Cleaning of Wounds

Saline or water is recommended to be used to irrigate the wound. High-pressure washing may be considered of greater significance than the solution used for washing.

5.4. Internal Contamination Intervention

When internal contamination occurs, addressing alpha- and beta-ray-emitting substances requires a complex intervention. The therapy aims to reduce absorption, enhance elimination, and block the distribution of the agent to target organs.

In the event of radioactive contamination from a nuclear accident, internal decontamination is used to remove radioactive materials from the bodies of those exposed. This process involves the use of certain drugs to prevent the uptake of radioactive substances into the body or to accelerate the removal of absorbed radionuclides. For example, stable iodine pills are used to prevent radioactive iodine from binding to the thyroid. It's important to note that chelating substances such as Prussian blue can decrease the absorption of radioactive isotopes like cesium and thallium from the intestines and increase their elimination from the body. The goal of internal decontamination is to reduce the risks associated with radiation exposure by preventing the accumulation of radioactive substances in the body. Internal decontamination should be carefully planned and controlled based on the type and level of exposure (Flynn and Goans, 2006).

6. Acute Radiation Syndrome (ARS) Treatment

Acute radiation syndrome (ARS) is caused by high levels of ionising radiation and is categorised into three main syndromes: hematopoietic, gastrointestinal, and neurovascular. In hematopoietic syndrome, the use of colony-stimulating factors (CSFs) (such as G-CSF and GM-CSF) has been proven to reduce the risk of infection by inducing the production of white blood cells, benefiting patients at risk of bone marrow failure with minimal risk. Gastrointestinal syndrome occurs following high radiation doses and usually requires intensive supportive care. This care includes treatment for severe dehydration, nausea, vomiting, and diarrhea, as well as enteral or parenteral nutrition support, and surveillance for sepsis and multiple organ failure. Neurovascular syndrome involves the central nervous system, and suggested treatment includes intensive supportive care such as brain perfusion maintenance and intracranial pressure management. Symptomatic treatment may involve steroids to decrease inflammation, anticonvulsants for seizure control, and sedatives for agitation. Successful management of these manifestations of ARS requires collaboration among several specialties and the use of intensive care resources, necessitating a synchronised process (Donnelly et al., 2010).

7. Local Radiation Injury

Localised radiation injuries (LRI) are caused by direct contact with high-energy radioactive sources. These injuries may begin as skin injuries but can advance to affect all tissues and organs of the body. Therefore, the management of LRI should be approached in the same way as burns. This may involve providing essential wound care, performing debridement, and carrying out amputation when it is deemed necessary. Examples of non-invasive diagnostic and therapeutic procedures include high-frequency ultrasound, thermography, capillary microscopy, dermatoscopy, and magnetic resonance imaging. Treatment is currently somewhat empirical and symptomatic as there is no standardised approach, and treatment is undertaken according to the extent of damage and the stage of lesions. There have been recent therapeutic approaches that show great promise, such as stem cell therapy, where human mesenchymal stem cells have shown promise in tissue reconstruction. These applications have an important place in a wide variety of fields such as tissue engineering and regenerative medicine. This has the potential to maximise the improvement of the management of radiation injuries and the healing process for the affected individuals (Müller and Meineke 2010).

8. Psychological Impacts

Fukushima not only demonstrated the immediate dangers people faced but also the significant psychological impact on the population. The psychological effects included post-traumatic stress from the explosion and evacuation. People lived in constant fear and guilt, especially parents worried about the potential impact of radiation on their children. There was also a sense of ambiguous loss, as evacuees were unsure if they could return home, leading to depressive symptoms. This separation forced families apart, reduced social cohesion, and increased conflicts and frustrations in the community. Additionally, evacuees experienced self-stigma, discrimination, and low self-esteem. These psychological effects highlight the importance of focusing on mental health care and services following nuclear disasters (A. Hasegawa et al., 2016).

Conclusion

The management of nuclear accidents requires a coordinated response from various medical and emergency disciplines. Given the potential for immediate and long-term consequences, an efficient response is critical for minimising adverse effects on human health and the environment. Preparedness for emergencies, including detailed contingency planning and communication strategies, is crucial. The triage systems described here are essential for the successful management of a nuclear accident.

Treatment of ARS and LRI shows the necessity of targeted medical therapies and the timely role of teamwork in the management of radiation emergencies. Modern medical technology integration, biodosimetry tools, and novel therapeutic approaches, for instance, are of tremendous importance for improving the medical results of patients.

In summary, further improvement of emergency preparedness and response, as well as enhancement of medical response and radiation protection, will substantially contribute to the enhanced resistance of health systems to nuclear accidents. Emergency response systems also need to be well-equipped for such accidents and then integrated with such development. This is important not only in the short term for managing such incidents but also in the long term for the provision of recovery and rehabilitation among the communities affected by such incidents. Past accidents should also serve as a guide for future preparedness techniques; the safety of the population should take priority over the challenges of nuclear accidents.

References

AFAD. (2024) "Nükleer Tehdit ve Tehlikeler." n.d. Accessed January 26, 2024. https://www.afad.gov.tr/kbrn/nukleer-tehditler.

Bazyar, J., Farrokhi, M., & Khankeh, H. (2019). Triage Systems in Mass Casualty Incidents and Disasters: A Review Study with A Worldwide Approach. *Open Access Macedonian Journal of Medical Sciences*, *7*(3), 482–494. https://doi.org/10.3889/oamjms.2019.119.

Bomanji, J. B., Novruzov, F., & Vinjamuri, S. (2014). Radiation accidents and their management: Emphasis on the role of nuclear medicine professionals. *Nuclear Medicine Communications*, *35*(10), 995–1002. https://doi.org/10.1097/MNM.0000000000000156.

Byrne, J., & Hoffman, S. M. (1987). Nuclear Power and Technological Authoritarianism. *Bulletin of Science, Technology & Society*, *7*(3–4), 658–671. https://doi.org/10.1177/027046768700700338.

Donnelly, E. H., Nemhauser, J. B., Smith, J. M., Kazzi, Z. N., Farfán, E. B., Chang, A. S., & Naeem, S. F. (2010). Acute Radiation Syndrome: Assessment and Management: *Southern Medical Journal*, *103*(6), 541–546. https://doi.org/10.1097/SMJ.0b013e3181ddd571.

Emer, M. Ö. (2017). Hospital Preparation in the Radiological/Nuclear Urgents and Organization of the Emergency Response Team. *Nuclear Medicine Seminars*, *3*(3), 225–230. https://doi.org/10.4274/nts.2017.028.

Flynn, D. F., & Goans, R. E. (2006). Nuclear Terrorism: Triage and Medical Management of Radiation and Combined-Injury Casualties. *Surgical Clinics of North America*, *86*(3), 601–636. https://doi.org/10.1016/j.suc.2006.03.005.

Hasegawa, A., Ohira, T., Maeda, M., Yasumura, S., & Tanigawa, K. (2016). Emergency Responses and Health Consequences after the Fukushima Accident; Evacuation and Relocation. *Clinical Oncology*, *28*(4), 237–244. https://doi.org/10.1016/j.clon.2016.01.002.

Hasegawa, A., Tanigawa, K., Ohtsuru, A., Yabe, H., Maeda, M., Shigemura, J., Ohira, T., Tominaga, T., Akashi, M., Hirohashi, N., Ishikawa, T., Kamiya, K., Shibuya, K., Yamashita, S., & Chhem, R. K. (2015). Health effects of radiation and other health problems in the aftermath of nuclear accidents, with an emphasis on Fukushima. *The Lancet*, *386*(9992), 479–488. https://doi.org/10.1016/S0140-6736(15)61106-0.

Hick, J. L., Bader, J. L., Coleman, C. N., Ansari, A. J., Chang, A., Salame-Alfie, A., Hanfling, D., & Koerner, J. F. (2018). Proposed "Exposure And Symptom Triage" (EAST) Tool to Assess Radiation Exposure After a Nuclear Detonation. *Disaster Medicine and Public Health Preparedness*, *12*(3), 386–395. https://doi.org/10.1017/dmp.2017.86.

Hrdina, C. M., Coleman, C. N., Bogucki, S., Bader, J. L., Hayhurst, R. E., Forsha, J. D., Marcozzi, D., Yeskey, K., & Knebel, A. R. (2009). The "RTR" Medical Response System for Nuclear and Radiological Mass-Casualty Incidents: A Functional TRiage-TReatment-TRansport Medical Response Model. *Prehospital and Disaster Medicine*, *24*(3), 167–178. https://doi.org/10.1017/S1049023X00006774.

Karaçalıoğlu, A. Ö. (2017). Medical Triage in Nuclear-Radiological Emergency. *Nuclear Medicine Seminars*, *3*(3), 196–200. https://doi.org/10.4274/nts.2017.022.

Maître, M., Croüail, P., Schneider, T., Kuroda, Y., Miyazaki, M., Tanigawa, K., Oughton, D., Tomkiv, Y., Skuterud, L., Liutsko, L., Charron, S., Pölz-Viol, C., Kesminiene, A., & Ostroumova, E. (2021). Living conditions and health status of populations living in territories impacted by nuclear accidents – Some lessons for developing health surveillance programme. *Environment International*, *147*, 106294. https://doi.org/10.1016/j.envint.2020.106294.

Morimura, N., Asari, Y., Yamaguchi, Y., Asanuma, K., Tase, C., Sakamoto, T., Aruga, T., & Members of the Japanese Association for Acute Medicine, Emergency, Task Force on the Fukushima Nuclear Power Plant Accident (JAAM-TF-FNPPA). (2013). Emergency/disaster medical support in the restoration project for the Fukushima nuclear power plant accident. *Emergency Medicine Journal, 30*(12), 997–1002. https://doi.org/10.1136/emermed-2012-201629.

Müller, K., & Meineke, V. (2010). Advances in the Management of Localized Radiation Injuries. *Health Physics, 98*(6), 843–850. https://doi.org/10.1097/HP.0b013e3181adcba7.

Nükleer Tehdit ve Tehlikeler. (n.d.). Retrieved 26 January 2024, from https://www.afad.gov.tr/kbrn/nukleer-tehditler.

Ohba, T., Liutsko, L., Schneider, T., Francesc Barquinero, J., Crouaïl, P., Fattibene, P., Kesminiene, A., Laurier, D., Sarukhan, A., Skuterud, L., Tanigawa, K., Tomkiv, Y., & Cardis, E. (2021). The SHAMISEN Project: Challenging historical recommendations for preparedness, response and surveillance of health and well-being in case of nuclear accidents: Lessons learnt from Chernobyl and Fukushima. *Environment International, 146*, 106200. https://doi.org/10.1016/ j.envint. 2020.106200.

Ohba, T., Tanigawa, K., & Liutsko, L. (2021). Evacuation after a nuclear accident: Critical reviews of past nuclear accidents and proposal for future planning. *Environment International, 148*, 106379. https://doi.org/10.1016/j.envint.2021.106379.

Ohtsuru, A., Tanigawa, K., Kumagai, A., Niwa, O., Takamura, N., Midorikawa, S., Nollet, K., Yamashita, S., Ohto, H., Chhem, R. K., & Clarke, M. (2015). Nuclear disasters and health: Lessons learned, challenges, and proposals. *The Lancet, 386*(9992), 489–497. https://doi.org/10.1016/S0140-6736(15)60994-1.

Ovalı, E. (n.d.). *Radyasyon Kazaları.* https://www.thd.org.tr/thddata/books /29/radyasyon-kazalari-ercument-ovali.pdf.

Paans, A. M. J., & De Jong, J. R. (2017). The Role of a Nuclear Medicine Department in Nuclear Accidents. In A. W. J. M. Glaudemans, J. Medema, A. K. Van Zanten, R. A. J. O. Dierckx, & C. T. B. Ahaus (Eds.), *Quality in Nuclear Medicine* (pp. 159–166). Springer International Publishing. https://doi.org/10.1007/978-3-319-33531-5_10.

Plants, C. on L. L. from the F. N. A. for I. S. and S. of U. S. N., Board, N. and R. S., Studies, D. on E. and L., & Council, N. R. (2014). Lessons Learned: Offsite Emergency Management. In *Lessons Learned from the Fukushima Nuclear Accident for Improving Safety of U.S. Nuclear Plants.* National Academies Press (US). https://www.ncbi.nlm.nih.gov/books/NBK253930/.

Ramesh, A., & Kumar, S. (2010). Triage, monitoring, and treatment of mass casualty events involving chemical, biological, radiological, or nuclear agents. *Journal of Pharmacy And Bioallied Sciences, 2*(3), 239. https://doi.org/10.4103/0975-7406.68506.

Schleipman, A., Gerbaudo, V., & Castronovo, F. (2004). Radiation disaster response: Preparation and simulation experience at an academic medical center. *Journal of Nuclear Medicine Technology* 32(1):22–27.

Sullivan, J. M., Prasanna, P. G. S., Grace, M. B., Wathen, L. K., Wallace, R. L., Koerner, J. F., & Coleman, C. N. (2013). Assessment of Biodosimetry Methods for a Mass-Casualty Radiological Incident: Medical Response and Management Considerations. *Health Physics, 105*(6), 540–554. https://doi.org/10.1097/HP.0b013e31829cf221.

Tintinalli, J. E., Ma, O. J., Yealy, D. M., Meckler, G. D., Stapczynski, J. S., Cline, D., & Thomas, S. (Eds.). (2020). *Tintinalli's emergency medicine: A Comprehensive Study Guide* (9. Edition). McGraw-Hill.

Walls, R. M., Hockberger, R. S., Gausche-Hill, M., & Rosen, P. (Eds.). (2023). *Rosen's emergency medicine: Concepts and clinical practice*. Elsevier.

World Nuclear Association. (2024) "Fukushima Daiichi Accident Accessed February 9, 2024. https://world-nuclear.org/focus/fukushima-daiichi-accident/fukushima-daiichi-accident-faq.aspx#.

Chapter 8

Neurological Effects of Radiation

Müge Ayanoğlu*

Department of Pediatric Neurology, Aydın Adnan Menderes University, Aydın, Türkiye

Abstract

Ionising radiation, emitted as electromagnetic waves and particles, poses significant risks encountered in daily life, occupations, and medical procedures, including accidental exposures. Cellular responses to radiation induce DNA damage and generate reactive free radicals, potentially leading to cell death or genomic instability, with implications such as secondary tumour formation. Despite evolving understanding, mechanisms underlying radiation's effects on the central nervous system (CNS) remain complex and incompletely elucidated. Evidence suggests radiation suppresses neurogenesis in the hippocampus by depleting neural progenitor cells and altering the microenvironment, exacerbating oxidative stress and inflammation, and causing endothelial cell death. Case reports and studies from radiation therapy and nuclear incidents inform the spectrum of neurological complications. Early complications like acute radiation encephalopathy present within two weeks, characterised by headaches and nausea, while early-delayed complications manifest with fatigue and cognitive deficits up to six months post-exposure. Late complications, such as radionecrosis and radiation-induced dementia, may appear months to years later, involving tissue necrosis or progressive cognitive decline. Secondary brain and spine tumors are also long-term risks. Foetal exposure to radiation, exemplified by historical incidents, underscores risks like microcephaly and neurological deficits. Effects vary based on gestational timing, with higher vulnerability during late organogenesis.

---

* Corresponding Author's Email: mugeayanoglu_05@hotmail.com.

In: Triage in Radiation Accidents
Editors: Nural Öztürk and Sibel Karaca
ISBN: 979-8-89530-415-0

In conclusion, insights into radiation-induced neurological disorders underscore the need for cautious therapeutic and diagnostic radiation use. Clinicians must balance these risks with benefits, emphasising safety measures in nuclear practices. This chapter highlights the importance of further research to comprehensively understand and reduce the impact of radiation on central nervous system health.

Keywords: encephalopathy, neurological effects, nuclear accident, radiation, radiotherapy

1. Introduction

Ionising radiation constitutes a form of energy emitted by atoms in the guise of electromagnetic waves and particles. Individuals can encounter radiation in public settings during routine activities, through occupational exposures, or via medical procedures (Organization, 1998). Accidental exposure incidents are also plausible. Ionising radiation can instigate DNA damage, comprising single-strand breaks, double-strand breaks, base damage, and DNA crosslinks, alongside the generation of highly reactive free radicals. While diverse DNA repair mechanisms may uphold genomic stability, certain lesions can prompt cell demise via apoptosis, autophagy, or mitotic catastrophe. Additionally, surviving cells may potentially engender late sequelae, such as the development of secondary tumours (Joiner and van der Kogel, 2018).

The underlying mechanisms elucidating the effects of radiation on the central nervous system (CNS) remain inadequately understood. Available evidence suggests that radiation may curtail neurogenesis in the hippocampus. Preclinical data indicate that radiation can deplete neural progenitor cells through multiple cell divisions. Furthermore, perturbations within the cellular microenvironment post-radiation exposure likely contribute significantly to diminished neurogenesis. Radiation-induced oxidative stress can activate proinflammatory pathways, leading to an upsurge in activated microglia, which may further impair neurogenesis. Additionally, radiation exposure can cause endothelial cell death, precipitating thrombus formation on the exposed matrix and infarction of small vessels (Wilke et al., 2018).

Case reports detailing complications post-radiotherapy and evidence from preclinical studies provide insights into the neurological effects of radiation on the nervous system. Moreover, findings from certain studies and implications drawn from catastrophic nuclear incidents and atomic bombings have augmented our understanding of these effects. Given the limited evidence from nuclear accidents, this review predominantly draws upon implications from case reports and studies following radiation therapy.

2. Neurological Complications of Radiation

Radiation exposure can impact the CNS with varying onset times for symptoms, categorized into early-onset, early-delayed, or late-onset complications. Symptoms appearing within 2 weeks, between 2 weeks to 6 months, and 6-12 months post-exposure are classified as early-onset, early-delayed, and late-onset complications, respectively. Early and early-delayed complications are generally reversible at least to some extent. However, late complications are irreversible (Tofilon and Fike, 2000). Brainstem, hippocampus, cranial nerves, spinal cord, cauda equina, brachial plexus, Circle of Willis, and peripheral nerves may be affected due to radiation exposure (Emami et al., 1991).

2.1. Early Complications

Acute radiation encephalopathy may manifest within the first two weeks following a radiotherapy session. Commonly reported symptoms include drowsiness, headache, nausea, and vomiting. Pre-existing neurological abnormalities may exacerbate following radiation due to disruption of the blood-brain barrier, resulting in elevated intracranial pressure (Brandsma et al., 2008; Dietrich et al., 2008; Young et al., 1974). Although sometimes magnetic resonance imaging (MRI) typically reveals no abnormalities, diffuse brain swelling may be observed (Brandsma et al., 2008).

2.2. Early-Delayed Complications

Symptoms emerge during the 2 weeks to 6 months post-exposure period. Patients may experience fatigue, drowsiness, lethargy, excessive daytime sleepiness, attention deficits, and memory impairments, similarly to patients with early complications. Additionally, somnolence syndrome, characterised by hypersomnolence, nausea, and anorexia, is often observed in children post-prophylactic cranial radiotherapy. Precise mechanisms are vasogenic edema and transient demyelination in the CNS (Tofilon and Fike, 2000). Magnetic resonance imaging results reveal a range of abnormalities, from hyperintense white matter lesions without enhancing on T2-weighted images to enhancing lesions around the irradiated area that were not previously present (Brandsma et al., 2008). Clinical evaluation, along with nonspecific findings on MRI and EEG, may necessitate corticosteroid therapy if symptoms persist (Faithfull and Brada, 1998).

Some patients may experience worsening neurological disorders approximately 6-12 weeks post-radiotherapy. In this case, two possibilities should be kept in mind: *i)* tumour progression, or *ii)* pseudoprogression. The mechanism of pseudoprogression may be explained by a better response of the tumour to radiotherapy. In addition to the symptoms of fatigue, drowsiness, lethargy, excessive daytime sleepiness, poor attention and memory deficits may be seen in patients with pseudoprogression up to 6 months after radiotherapy, which may not indicate a long-term cognitive decline (Brandes et al., 2008; Brandsma et al., 2008; Dietrich et al., 2008). It is difficult to distinguish between two clinical pictures, since MRI results may indicate a larger size of the tumour, even in the presence of pseudoprogression.

Brainstem syndrome is a rare complication that may occur in patients receiving radiotherapy for head or neck cancers, where targeted fields overlap with the brainstem. Clinical symptoms may include ataxia, dysarthria, hearing loss, diplopia, and nystagmus. However, some patients may be diagnosed with the findings on MRI without any clinical symptoms. Radiological findings may reveal hyperintensities in the white matter of the brainstem and cerebellum with or without enhancement (Kosmin and Rees, 2022). In a study by Devine et al., 107 pediatric patients were included who received photon radiation. Among 107 patients, 2 of them developed grade 1 brainstem necrosis without any clinical symptoms. Moreover, MRI findings were resolved spontaneously (Devine et al., 2019). Although corticosteroids may benefit some patients, severe cases may rarely progress to coma or death.

2.3. Late Complications

Neurological symptoms appearing after 6-12 months following radiotherapy should raise suspicion of late complications of radiation. There are two main clinical pictures of late complications: *i)* radionecrosis, and *ii)* late-delayed radiation encephalopathy. Radionecrosis may occur after focal radiotherapy for brain metastasis or due to overlapping brain fields in cases of extraaxial and extracranial tumours. Additionally, patients who have undergone stereotactic surgery for both neoplastic and non-neoplastic conditions such as arterio-venous malformations may develop radionecrosis. Pathological features of radionecrosis include fibrinoid changes in vessels, coagulative necrosis, demyelination, and gliosis, which may be explained by the release of vascular endothelial growth factor (VEGF) in response to hypoxia (Tofilon and Fike, 2000). In the presence of radionecrosis, MRI findings may show Swiss cheese-like enhancing pattern, as enhancing lesions may include a central non-enhancing pattern due to necrosis. The white matter of the brain is vulnerable to develop radionecrosis due to relatively poor blood supply. By the time, the necrotic areas may be replaced with atrophy. Perfusion MRI and MR spectroscopy may be helpful in the differential diagnosis of these clinical pictures. However, sometimes histopathological examination may be required. Patients' symptoms may ameliorateto dexamethasone and surgical resection of necrotic foci. Additionally, hyperbaric oxygen therapy, pentoxifylline, alpha-tocopherol, and pentobarbital may also be considered. Recently, it has been proposed that patients may benefit from anti-VEGF monoclonal antibody, known as bevacizumab (Deibert et al., 2013). However, repeated doses of bevacizumab, or surgical resection may be needed, especially in case of a diagnostic uncertainity.

2.3.1. Effects of Radiation on Cognitive Funcitions

Cognitive decline and dementia may ensue 6-12 months post-radiotherapy, involving chronic damage to progenitor cell populations crucial for hippocampal neurogenesis and white matter integrity (Dietrich et al., 2008). While clinical symptoms remain stable in most cases, a slowly progressive course may also be seen. Although no curative treatment is available, anticholinesterases may alleviate symptoms (Cramer et al., 2019; Crossen et al., 1994; Lawrie et al., 2019).

Radiation-induced dementia presents as diffuse white matter injury, which may manifest 2 years after exposure. Initial symptoms may resemble normal pressure hydrocephalus, with developing progressive memory loss, emotional lability, apathy, attention deficit, fatigue, urinary incontinence, and gait difficulties. Radiological findings indicate ventricular enlargement, cortical and subcortical atrophy. Although specific medication is not available, ventriculo-peritoneal shunting may provide short-term improvement. Factors such as female gender, younger age at the treatment, high dose radiotherapy, concurrent chemotherapy, cerebellar mutism, and a history of stroke or seizures are associated with a higher risk of cognitive decline in children who underwent radiotherapy (Armstrong et al., 2011; Di et al., 2012; Orgel et al., 2016). Decreased intellectual quotient and academic performance may be more frequently seen, especially if the hippocampi and temporal lobes are affected. The degree of white matter loss correlates with the decline in cognitive functions (Jacola et al., 2014; Palmer et al., 2001; Reddick et al., 2003).

2.3.1.1. Radiation-Related Secondary Tumours

Patients with a history of radiotherapy have a tendency to develop brain and spine tumours such as meningiomas, gliomas, and sarcomas after the treatment (Taylor et al., 2010). Adult survivors who underwent cranial radiotherapy due to childhood acute lymphoblastic leukemia have a higher relative risk (Neglia et al., 1991; Ron et al., 1988). Second tumours may occur both early after the radiotherapy or after decades. Usually, early second tumours are gliomas or meningiomas. Most meningiomas can be successfully treated (Broniscer et al., 2004).

2.3.1.2. Radiation-related Vasculopathy

Radiation exposure may induce vasculopathy, accelerating atherosclerosis, particularly following neck and head radiotherapy. Microscopic findings reveal fibrous tissue replacing normal intima and media due to damaged internal elastic lamina. In cases of stroke, consideration of radiation-induced vasculopathy alongside other risk factors is important. Young children who experienced cranial radiotherapy may have a risk of developing Moyamoya disease, described as progressive occlusive cranial vasculopathy (Bitzer and Topka, 1995). In a study by Kralik et al., among 75 patients who underwent radiation therapy due to primary brain tumours, 6.7% of them developed steno-occlusive vasculopathy (Kralik et al., 2017). In another study, it was reported that 19% of 32 pediatric patients who underwent radiation therapy developed steno-occlusive vasculopathy (Omura, Aida, Sekido, Kakehi and Matsubara, 1997). In a review article, the estimated rate of Moyamoya disease after cranial radiotherapy was suggested as 3.5%. Especially, in the presence of a history of radiation therapy for suprasellar lesions, the major and the terminal branches of internal carotid arter are vulnerable to develop vasculopathy (Ullrich et al., 2007). Clinical symptoms include transient ischemic attacks, cerebrovascular infarcts, and seizures. Pathological examinations reveal myointimal proliferation and accumulation of foam cells (Aoki et al., 2002). Surgical interventions may help to prevent stroke attacks in these cases (Kralik et al., 2017; Smith and Scott, 2010). Patients with a history of neurofibromatosis type 1, exposure to radiation at a younger age, concurrent combined chemotherapy have higher risks of Moyamoya Disease (Ullrich et al., 2007).

In addition to steno-occlusive vasculopathies, vascular malformations including cavernomas, telangiectasias, and rarely aneurysms may be seen after cranial radiotherapy (Becker et al., 2021; Bitzer and Topka, 1995; Nanney III et al., 2014; Ullrich et al., 2007). The mechanisms of radiation-related aneurysms are not well understood. Precisely, it may be explained by the damage to endothelial cells, elastic lamina, and smooth mussels of the vessels. Also, the acceleration of the atherosclerosis process may have a contribution (Nanney III et al., 2014). Although aneurysms after radiotherapy are seen infrequently, they may have a fatal course. Moreover, there is an increased tendency to be ruptured in the presence of radiation-related aneurysms. Aneurysm development usually occurs approximately 10 years or more after exposure to radiation. However, in patients with a history of stereotactic surgery, the elapsed time from exposure to radiation until the development of aneurysms is shorter due to the use of high-energy focused radiation (Nanney III et al., 2014).

Cavernous malformations may develop after radiotherapy, particularly more frequently in children than in adults. Studies have shown that patients who underwent radiation therapy at younger than 10 years of age might develop cavernomas with a higher likelihood and in a shorter duration compared to other age groups. Also, there is a positive correlation between higher doses of radiation therapy and the likelihood of developing cavernomas (Heckl, Aschoff and Kunze, 2002). Exposure to radiation leads to ischemia of the vessel walls due to endothelial proliferation and fibrinoid necrosis. Subsequently, the production of hypoxia-inducible factor 1 and VEGF are activated which may lead to reactive and irregular angiogenesis. Although some patients may present with spontaneous intracranial hemorrhage, some of them are diagnosed incidentally (Cutsforth-Gregory, et al., 2015).

Mineralising microangiopathy with dystrophic calcification may occur after radiation therapy. Patients are usually asymptomatic and are diagnosed incidentally. Microscopic examinations of the lesions reveal microangiopathies with calcifications in the walls of pericapillary small vessels, probably due to injury secondary to radiation. The rate is higher in patients who underwent radiation therapy at a younger age than 10 years. Computed tomography reveals calcifications in the basal ganglia, subcortical white matter, cerebellum, and brainstem (Shanley, 1995; Valk and Dillon, 1991).

Patients may present with migraine-like headaches, seizures, and focal neurological deficits such as aphasia, neglect, aphasia, and hemiparesis, indicative of cortical dysfunction. This constellation of symptoms is termed stroke-like migraine attacks after radiotherapy (SMART syndrome). After a few weeks, many patients' stroke-like attacks may resolve spontaneously. Although the underlying mechanism is not well understood, it has been suggested that exposure to radiation might result in vasculary endothelial damage (Black et al., 2013). Radiological findings typically reveal pial enhancement and cortical high signal changes. Additionally, diffusion restriction and cortical oedema may be seen in various cases. The clinical course may result in permanent sequelae in some patients (Black et al., 2013; Kerklaan et al., 2011; Khanipour Roshan et al., 2015). Although treatment options are limited, isolated reports suggest potential benefits from L-arginine (Black et al., 2013; Dominguez and Malani, 2021; Khan et al., 2018).

2.3.1.3. Radiation-Related Myelopathy

Exposure to radiation may lead to complications in the spinal cord, including early and late-delayed myelopathy, lower motor neuron disorders, and spinal hemorrhage. Early myelopathy is a non-progressive clinical picture and may present with Lhermitte's phenomenon. The possible mechanism involves transient demyelination and oligodendroglial cell loss in the posterior columns.

Late-delayed myelopathy may occur 1-10 years after radiotherapy, especially in older patients, presenting with slowly progressive sensorimotor deficits, bladder dysfunction, and paresis. Radiological findings may show edema, atrophy, and cystic changes in the spinal cord. Microscopic findings indicate demyelination, focal necrosis, axonal loss, perivascular necrosis, and vasculitis. Magnetic resonance imaging reveals expansion and oedema of the cord and hyperintensities on T2-weighted images. Enhancement and hemorrhagic lesions may exist (Khan et al., 2018; Wong, Fehlings, & Sahgal, 2015).

Distal spinal cord and cauda equina may be affected after radiation therapy, known as lower motor neuron syndrome. Patients present with progressive weakness, muscle fasciculations without sensation abnormality. Sphincter function and sensation may be disrupted in later stages (Abraham and Drory, 2013). Although corticosteroids may be beneficial initially, neurological findings may progress to permanent sequelae. Hyperbaric oxygen and bevacizumab may be given in some selected cases (Khan et al., 2018).

2.3.1.4. Radiation-Related Cranial Nerve Dysfunction

Cranial nerve dysfunctions may occur as a result of radiation toxicity. Pathological examinations reveal disruption of the nerve-blood barrier, demyelination, ischaemia, fibrosis, and coagulation necrosis. Post-radiation cranial nerve palsy is known as a rare complication and usually presents a long latent period after radiation therapy (Teo et al., 2000). In a crossectional study by Kong et al., the rate and risk factors of cranial nerve palsy were investigated after radiation therapy due to nasopharymgeal cancer. The median follow-up period was 11.4 years, 30.9% of the 317 patients developed cranial nerve palsy with a median latent period of 7.6 years. The independent risk factors for developing cranial nerve palsy were; receiving chemotherapy, higher doses of radiation, and upper neck fibrosis (Kong et al., 2011).

A few case reports and case series exist regarding the radiological findings of radiation-induced cranial nerve damage. Smooth thickening with enhancement and subsequent atrophy may be seen on MRI (Archer, Liao and Trobe, 2019; Kong et al., 2011). Optic neuropathy should be considered if a patient with a history of radiation therapy presents with subacute painless vision loss. Ophthalmological examination may show disc swelling with surrounding exudate, hemorrhages, and subretinal fluid (Brown et al., 1982). Radiation-related optic neuropathy is particularly relevant for patients who underwent whole cranial radiotherapy or had treatment fields near the optic nerves and chiasma, including the cavernous sinus, orbit, pituitary gland, nasal cavity, or nasopharynx. It was proposed that exposure to radiation might be tolerated in doses less than 8 Gy for the optic nerves and chiasma. However, in some studies doses less than 10 Gy may be tolerated for the optic nerves (Girkin et al., 1997; Leber, Berglöff and Pendl, 1998; Ove et al., 2000). In a study by Stafford et al., the rate of radiation-related optic neuropathy was inverstigated among patients who underwent stereotactic radiosurgery. The results showed that less than 2% of the study group was diagnosed with radiation-related optic neuropathy, even majority of the group received more than 8 Gy (Stafford et al., 2003). Althoughno effective treatment is available, corticosteroids may be useful in early stages. Additionally, in case reports, some favourable results were reported with some treatments such as with pentoxifylline, bevacizumab, and with a combination of vitamin E and pentoxifylline (Chahal, Lam and Khaderi, 2013; Farooq et al., 2012; Gupta et al., 2001).

Other cranial nerves including abducens, trigeminal, facial, vestibulocochlear, glossopharyngeal, vagus, spinal accessory, and hypoglossal nerves may be affected after radiotherapy (Archer et al., 2019; Saremi et al., 2005; Vaphiades et al., 2011). In case of oculomotor, abducens, and trochlear nerve palsies, patients may present with diplopia and ophthalmoplegia. Moreover, pitosis may occur if oculomotor nerve palsy presents. In a case report, it was reported that post-radiation oculomotor nerve palsy may occur even after 2 decades (Grassmeyer et al., 2021).

Radiation and chemotherapy may lead to sensorineural hearing loss due to the impacts on organ of Corti of the cochlea and cochlear nerve (Mujica-Mota et al., 2014; Strojan et al., 2017). Additionally, as a result of injury of the spiral ganglion, radiation may cause abnormal auditory brainstem responses (Mujica-Mota et al., 2014). Receiving concurrent chemotherapy, especially cisplatin, and consumptions of alcohol and tobacco may deteriorate the process of radiation-related sensorineural hearing loss (Zuur et al., 2007). In case of a sudden sensorineural hearing loss, intratymphanic or oral steroids may be effective (Spear and Schwartz, 2011). There are some experimental studies investigating how to ameliorate the radiation-related ototoxicity with some drugs including L-N-Acetylcysteine, green tea extract, piracetam, metfromine, lipoic acid, brain-derived neurotrophic factor (BDNF) (W. K. Low et al., 2008; W. K. Low et al., 2006; Sheth et al., 2017; Shorter et al., 2014). However, the agents have rarely been investigated in clinical trials.

Swallowing, chewing, and speech functions may be injured after exposure to radiation. The difficulties may be due to fibrosis of the muscles that play a key role in swallowing and chewing, radiation-related injury of the cranial nerves responsible for swallowing and speech (IXth and X^{th}) and chewing (V^{th}), and injury to the medulla oblongata (Langmore and Krisciunas, 2010; Manikantan et al., 2009; Murphy and Gilbert, 2009; Platteaux eta., 2010). Symptoms emerging 0-3 months and 1 year after the exposure to radiation were described as acute dysphagia, and chronic dysphagia, respectively. Usually, the symptoms of acute dysphagia improve 6-12 months later (Heijnen et al., 2016; Murphy and Gilbert, 2009). Increased age, consumption of tobacco, the doses of radiotherapy and the proximity of the fields to the pharyngeal are the risk factors for developing swallowing, speech and chewing difficulty (Murphy and Gilbert, 2009; Platteaux et al., 2010). Interventions for rehabilitation such as neuromuscular electrical stimulation and balloon dilatation, exercises and manoeuvres, acupuncture, and postural techniques may ameloriate the symptoms (Long and Wu, 2013). Additionally, it was proposed that steroids, proton pump inhibitors and TNF-alpha may be useful (Villari and Courey, 2015). Increasing the hydration of the larynx with consumption of water, and upregulation of salivary glands with Substance P, bombesin and enkefalin have been proposed to relieve voice impairment (Lazarus, 2009; Lidegran et al., 1995; Villari and Courey, 2015).

2.3.1.5. Radiation-Related Pelxopathy

Plexopathy is a peripheral nerve complication after radiation therapy. Patients may present with pain, weakness and paralysis in the upper extremity after radiation therapy which is described as radiation-related plexopathy. Receiving concurrent chemotherapy and higher doses are related to higher risks (Chen et al., 2012; Metcalfe and Etiz, 2016; Platteaux et al., 2010). Patients may benefit from corticosteroids in early delayed plexopathy cases. However, a progressive clinical course and permanent neurological deficits are seen in later delayed plexopathy (Yamanaka and Hayano, 2017). In severe cases, surgical interventions including neurolysis, nerve grafting, and removal of the fibrotic tissue should be considered (Sakellariou et al., 2014).

Neuropathic pain may also be seen in addition to motor and sensory nerve dysfunctions in cases of cranial nerve neuropathies or plexopathy. Patients may benefit from corticosteroids, non-steroidal anti-inflammatory drugs. In unresponsive cases, opioids, gabapentin, and tyicyclic anti-depressants should be considered (Delanian, Lefaix and Pradat, 2012; Sakellariou et al., 2014).

3. Impacts of Radiation on Electroencephalography Patterns

In a prospective study performed 3-5 years and 10-13 years after the Chernobyl accident, conventional EEG results were assessed in patients who experienced acute radiation sickness. Control groups consisted of veterans of the Afghanistan war with similar neuropsychiatric problems including posttraumatic stress disorder, traumatic brain injury, and patients with dyscirculatory encephalopathy. The results showed changes compatible with cortical-limbic overactivation in the "3-5 years after irradiation" group. Moreover, EEG patterns in the "10-13 years after irradiation" group transformed towards patterns compatible with organic brain damage, with inhibition of the cortical-limbic system (Loganovsky and Yuryev, 2001).

4. Neurological Effects of Radition in the Case of Foetal Exposure

Pregnant women may be exposed to radiation accidentally or due to medical interventions. Reports regarding survivors of the Japanese atomic bomb and Chernobyl showed that radioactive fallout may pass through the placenta. Many reports and investigations about foetal exposure to radiation have focused on radiation-related thyroid toxicity, increased rates of leukemia, and solid tumours (Sugiyama et al., 2021). Although data are limited, babies exposed to radiation in utero may have smaller head sizes compared to those not exposed (Hatch et al., 2017). Additionally, microcephaly, nervous system abnormalities, abnormal behavior patterns, and intellectual disability have been reported in children exposed to radiation in utero (Radiation, 1988). Microcephaly may occur with or without intellectual disability (Schull and Otake, 1999). The possibility of neurological abnormalities increases at higher doses of exposure. Some neurological abnormalities such as brain heterotopia, neuronal depletion, abnormal synaptic organization, seizures, behavioral problems, and worse academic performance may occur in case of in utero exposure to radiation during late organogenesis (8-15 weeks post-conception). The vulnerability of cells due to high neuronal mitotic rate and migration in late organogenesis may explain this phenomenon. No evidence suggests higher rates of neurological outcomes with exposure to radiation before 8th weeks or after 26th weeks (Otake and Schull, 1998; Schull and Otake, 1999).

Conclusion

In conclusion, the comprehensive examination of radiation-induced neurological effects reveals a spectrum of complex and often debilitating conditions that can manifest across varying timelines post-exposure. From acute complications such as radiation encephalopathy to chronic conditions like radiation-induced dementia and secondary malignancies, the impact on the central nervous system underscores the importance of vigilance in medical practices involving ionising radiation. Insights gained from both clinical cases and epidemiological studies illuminate the intricate mechanisms through which radiation disrupts neurogenesis, triggers inflammatory responses, and leads to vascular pathologies. As we navigate advancements in therapeutic radiology and diagnostic imaging, it becomes increasingly crucial for healthcare providers to integrate robust safety protocols and radiation mitigation strategies. By doing so, we can mitigate the risks associated with radiation exposure and strive to optimize patient outcomes while safeguarding neurological health in both clinical and environmental settings.

References

Abraham, A., & Drory, V. (2013). Postradiation lower motor neuron syndrome: case series and literature review. *Journal of neurology, 260*, 1802-1806.

Aoki, S., Hayashi, N., Abe, O., Shirouzu, I., Ishigame, K., Okubo, T., . . . Araki, T. (2002). Radiation-induced arteritis: thickened wall with prominent enhancement on cranial MR images—report of five cases and comparison with 18 cases of moyamoya disease. *Radiology, 223*(3), 683-688.

Archer, E. L., Liao, E. A., & Trobe, J. D. (2019). Radiation-induced optic neuropathy: clinical and imaging profile of twelve patients. *Journal of Neuro-Ophthalmology, 39*(2), 170-180.

Armstrong, G. T., Conklin, H. M., Huang, S., Srivastava, D., Sanford, R., Ellison, D. W., Robison, L. L. (2011). Survival and long-term health and cognitive outcomes after low-grade glioma. *Neuro-oncology, 13*(2), 223-234.

Becker, L., Gebauer, J., Küchler, J., Staackmann, C., Schacht, H., Lauten, M., Neumann, A. (2021). Are radiation-induced cavernomas clinically relevant findings? Results from long-term follow-up with brain magnetic resonance imaging of childhood cancer survivors. *Radiology and Oncology, 55*(3), 274-283.

Bitzer, M., & Topka, H. (1995). Progressive cerebral occlusive disease after radiation therapy. *Stroke, 26*(1), 131-136.

Black, D. F., Morris, J., Lindell, E., Krecke, K., Worrell, G., Bartleson, J., & Lachance, D. (2013). Stroke-like migraine attacks after radiation therapy (SMART) syndrome is not always completely reversible: a case series. *American Journal of Neuroradiology, 34*(12), 2298-2303.

Brandes, A. A., Tosoni, A., Spagnolli, F., Frezza, G., Leonardi, M., Calbucci, F., & Franceschi, E. (2008). Disease progression or pseudoprogression after concomitant radiochemotherapy treatment: pitfalls in neurooncology. *Neuro-oncology, 10*(3), 361-367.

Brandsma, D., Stalpers, L., Taal, W., Sminia, P., & van den Bent, M. J. (2008). Clinical features, mechanisms, and management of pseudoprogression in malignant gliomas. *The lancet oncology, 9*(5), 453-461.

Broniscer, A., Ke, W., Fuller, C. E., Wu, J., Gajjar, A., & Kun, L. E. (2004). Second neoplasms in pediatric patients with primary central nervous system tumors: the St. Jude Children's Research Hospital experience. *Cancer: Interdisciplinary International Journal of the American Cancer Society, 100*(10), 2246-2252.

Brown, G. C., Shields, J. A., Sanborn, G., Augsburger, J. J., Savino, P. J., & Schatz, N. J. (1982). Radiation optic neuropathy. *Ophthalmology, 89*(12), 1489-1493.

Chahal, H. S., Lam, A., & Khaderi, S. K. (2013). Is pentoxifylline plus vitamin E an effective treatment for radiation-induced optic neuropathy? *Journal of Neuro-Ophthalmology, 33*(1), 91-93.

Chen, A. M., Hall, W. H., Li, J., Beckett, L., Farwell, D. G., Lau, D. H., & Purdy, J. A. (2012). Brachial plexus-associated neuropathy after high-dose radiation therapy for head-and-neck cancer. *International Journal of Radiation Oncology* Biology* Physics, 84*(1), 165-169.

Cramer, C. K., Cummings, T. L., Andrews, R. N., Strowd, R., Rapp, S. R., Shaw, E. G., Lesser, G. J. (2019). Treatment of radiation-induced cognitive decline in adult brain tumor patients. *Current treatment options in oncology, 20*, 1-16.

Crossen, J. R., Garwood, D., Glatstein, E., & Neuwelt, E. A. (1994). Neurobehavioral sequelae of cranial irradiation in adults: a review of radiation-induced encephalopathy. *Journal of Clinical Oncology, 12*(3), 627-642.

Cutsforth-Gregory, J. K., Lanzino, G., Link, M. J., Brown, R. D., & Flemming, K. D. (2015). Characterization of radiation-induced cavernous malformations and comparison with a nonradiation cavernous malformation cohort. *Journal of neurosurgery, 122*(5), 1214-1222.

Deibert, C. P., Ahluwalia, M. S., Sheehan, J. P., Link, M. J., Hasegawa, T., Yomo, S., Lunsford, L. D. (2013). Bevacizumab for refractory adverse radiation effects after stereotactic radiosurgery. *Journal of neuro-oncology, 115*, 217-223.

Delanian, S., Lefaix, J.-L., & Pradat, P.-F. (2012). Radiation-induced neuropathy in cancer survivors. *Radiotherapy and Oncology, 105*(3), 273-282.

Devine, C. A., Liu, K. X., Ioakeim-Ioannidou, M., Susko, M., Poussaint, T. Y., Huisman, T. A., Rao, A. D. (2019). Brainstem injury in pediatric patients receiving posterior fossa photon radiation. *International Journal of Radiation Oncology* Biology* Physics, 105*(5), 1034-1042.

Di Pinto, M., Conklin, H. M., Li, C., & Merchant, T. E. (2012). Learning and memory following conformal radiation therapy for pediatric craniopharyngioma and low-grade glioma. *International Journal of Radiation Oncology* Biology* Physics, 84*(3), e363-e369.

Dietrich, J., Monje, M., Wefel, J., & Meyers, C. (2008). Clinical patterns and biological correlates of cognitive dysfunction associated with cancer therapy. *The oncologist, 13*(12), 1285-1295.

Dominguez, M., & Malani, R. (2021). Stroke-like migraine attacks after radiation therapy (SMART) syndrome: a comprehensive review. *Current Pain and Headache Reports, 25*, 1-9.

Emami, B., Lyman, J., Brown, A., Cola, L., Goitein, M., Munzenrider, J., . . . Wesson, M. (1991). Tolerance of normal tissue to therapeutic irradiation. *International Journal of Radiation Oncology* Biology* Physics, 21*(1), 109-122.

Faithfull, S., & Brada, M. (1998). Somnolence syndrome in adults following cranial irradiation for primary brain tumours. *Clinical Oncology, 10*(4), 250-254.

Farooq, O., Lincoff, N. S., Saikali, N., Prasad, D., Miletich, R. S., & Mechtler, L. L. (2012). Novel treatment for radiation optic neuropathy with intravenous bevacizumab. *Journal of Neuro-Ophthalmology, 32*(4), 321-324.

Girkin, C. A., Comey, C. H., Lunsford, L. D., Goodman, M. L., & Kline, L. B. (1997). Radiation optic neuropathy after stereotactic radiosurgery. *Ophthalmology, 104*(10), 1634-1643.

Grassmeyer, J. J., Fernandes, J. A., Helvey, J. T., & Kedar, S. (2021). Radiation-induced bilateral oculomotor nerve palsy 20 years after radiation treatment. *Neurology, 96*(20), 955-957.

Gupta, P., Meisenberg, B., Amin, P., & Pomeranz, H. D. (2001). Radiation Retinopathy: The Role of Pentoxifylline. *Retina, 21*(5), 545-547.

Hatch, M., Little, M. P., Brenner, A. V., Cahoon, E. K., Tereshchenko, V., Chaikovska, L., Shpak, V. (2017). Neonatal outcomes following exposure in utero to fallout from Chernobyl. *European journal of epidemiology, 32*, 1075-1088.

Heckl, S., Aschoff, A., & Kunze, S. (2002). Radiation-induced cavernous hemangiomas of the brain: A late effect predominantly in children. *Cancer, 94*(12), 3285-3291.

Heijnen, B., Speyer, R., Kertscher, B., Cordier, R., Koetsenruijter, K., Swan, K., & Bogaardt, H. (2016). Dysphagia, speech, voice, and trismus following radiotherapy and/or chemotherapy in patients with head and neck carcinoma: review of the literature. *BioMed research international, 2016*(1), 6086894.

Jacola, L. M., Ashford, J. M., Reddick, W. E., Glass, J. O., Ogg, R. J., Merchant, T. E., & Conklin, H. M. (2014). The relationship between working memory and cerebral white matter volume in survivors of childhood brain tumors treated with conformal radiation therapy. *Journal of neuro-oncology, 119*, 197-205.

Joiner, M. C., & van der Kogel, A. J. (2018). *Basic clinical radiobiology*: CRC press.

Kerklaan, J. P., Lycklama á Nijeholt, G. J., Wiggenraad, R. G., Berghuis, B., Postma, T. J., & Taphoorn, M. J. (2011). SMART syndrome: a late reversible complication after radiation therapy for brain tumours. *Journal of neurology, 258*, 1098-1104.

Khan, M., Ambady, P., Kimbrough, D., Shoemaker, T., Terezakis, S., Blakeley, J., Izbudak, I. (2018). Radiation-induced myelitis: initial and follow-up MRI and clinical features in patients at a single tertiary care institution during 20 years. *American Journal of Neuroradiology, 39*(8), 1576-1581.

Khanipour Roshan, S., Salmela, M. B., & McKinney, A. M. (2015). Susceptibility-weighted imaging in stroke-like migraine attacks after radiation therapy syndrome. *Neuroradiology, 57*, 1103-1109.

Kong, L., Lu, J. J., Liss, A. L., Hu, C., Guo, X., Wu, Y., & Zhang, Y. (2011). Radiation-induced cranial nerve palsy: a cross-sectional study of nasopharyngeal cancer patients after definitive radiotherapy. *International Journal of Radiation Oncology* Biology* Physics, 79*(5), 1421-1427.

Kosmin, M., & Rees, J. (2022). Radiation and the nervous system. *Practical Neurology, 22*(6), 450-460.

Kralik, S. F., Watson, G. A., Shih, C.-S., Ho, C. Y., Finke, W., & Buchsbaum, J. (2017). Radiation-induced large vessel cerebral vasculopathy in pediatric patients with brain tumors treated with proton radiation therapy. *International Journal of Radiation Oncology* Biology* Physics, 99*(4), 817-824.

Langmore, S. E., & Krisciunas, G. P. (2010). Dysphagia after radiotherapy for head and neck cancer: etiology, clinical presentation, and efficacy of current treatments. *Perspectives on Swallowing and Swallowing Disorders (Dysphagia), 19*(2), 32-38.

Lawrie, T. A., Gillespie, D., Dowswell, T., Evans, J., Erridge, S., Vale, L., . . . Grant, R. (2019). Long-term neurocognitive and other side effects of radiotherapy, with or without chemotherapy, for glioma. *Cochrane Database of Systematic Reviews* (8).

Lazarus, C. L. (2009). Effects of chemoradiotherapy on voice and swallowing. *Current opinion in otolaryngology & head and neck surgery, 17*(3), 172-178.

Leber, K. A., Berglöff, J., & Pendl, G. (1998). Dose—response tolerance of the visual pathways and cranial nerves of the cavernous sinus to stereotactic radiosurgery. *Journal of neurosurgery, 88*(1), 43-50.

Lidegran, M., Domeij, S., Dahlqvist, Å., Henriksson, R., Franzén, L., Gustafsson, H., & Forsgren, S. (1995). Irradiation influences the expression of substance P and enkephalin in the rat larynx. *Cell and tissue research, 279*, 55-63.

Loganovsky, K. N., & Yuryev, K. L. (2001). EEG patterns in persons exposed to ionizing radiation as a result of the Chernobyl accident: part 1: conventional EEG analysis. *The Journal of neuropsychiatry and clinical neurosciences, 13*(4), 441-458.

Long, Y.-B., & Wu, X.-P. (2013). A randomized controlled trail of combination therapy of neuromuscular electrical stimulation and balloon dilatation in the treatment of radiation-induced dysphagia in nasopharyngeal carcinoma patients. *Disability and rehabilitation, 35*(6), 450-454.

Low, W.-K., Sun, L., Tan, M. G., Chua, A. W., & Wang, D.-Y. (2008). LN-Acetylcysteine protects against radiation-induced apoptosis in a cochlear cell line. *Acta oto-laryngologica, 128*(4), 440-445.

Low, W. K., Toh, S. T., Wee, J., Fook-Chong, S. M., & Wang, D. Y. (2006). Sensorineural hearing loss after radiotherapy and chemoradiotherapy: a single, blinded, randomized study. *Journal of Clinical Oncology, 24*(12), 1904-1909.

Manikantan, K., Khode, S., Sayed, S. I., Roe, J., Nutting, C. M., Rhys-Evans, P., Kazi, R. (2009). Dysphagia in head and neck cancer. *Cancer treatment reviews, 35*(8), 724-732.

Metcalfe, E., & Etiz, D. (2016). Early transient radiation-induced brachial plexopathy in locally advanced head and neck cancer. *Contemporary Oncology/Współczesna Onkologia, 20*(1), 67-72.

Mujica-Mota, M. A., Lehnert, S., Devic, S., Gasbarrino, K., & Daniel, S. J. (2014). Mechanisms of radiation-induced sensorineural hearing loss and radioprotection. *Hearing research, 312*, 60-68.

Murphy, B. A., & Gilbert, J. (2009). *Dysphagia in head and neck cancer patients treated with radiation: assessment, sequelae, and rehabilitation.* Paper presented at the Seminars in radiation oncology.

Nanney III, A. D., El Tecle, N. E., El Ahmadieh, T. Y., Daou, M. R., Ivan, E. N. B., Marymont, M. H., Bendok, B. R. (2014). Intracranial aneurysms in previously irradiated fields: literature review and case report. *World neurosurgery, 81*(3-4), 511-519.

Neglia, J. P., Meadows, A. T., Robison, L. L., Kim, T. H., Newton, W. A., Ruymann, F. B., Hammond, G. D. (1991). Second neoplasms after acute lymphoblastic leukemia in childhood. *New England Journal of Medicine, 325*(19), 1330-1336.

Omura, M., Aida, N., Sekido, K., Kakehi, M., & Matsubara, S. (1997). Large intracranial vessel occlusive vasculopathy after radiation therapy in children: clinical features and usefulness of magnetic resonance imaging. *International Journal of Radiation Oncology* Biology* Physics, 38*(2), 241-249.

Organization, W. H. (1998). Diagnosis and treatment of radiation injuries.

Orgel, E., O'Neil, S. H., Kayser, K., Smith, B., Softley, T. L., Sherman-Bien, S., Freyer, D. R. (2016). Effect of sensorineural hearing loss on neurocognitive functioning in pediatric brain tumor survivors. *Pediatric blood & cancer, 63*(3), 527-534.

Otake, M., & Schull, W. J. (1998). Radiation-related brain damage and growth retardation among the prenatally exposed atomic bomb survivors. *International journal of radiation biology, 74*(2), 159-171.

Ove, R., Kelman, S., Amin, P. P., & Chin, L. S. (2000). Preservation of visual fields after peri-sellar gamma-knife radiosurgery. *International journal of cancer, 90*(6), 343-350.

Palmer, S. L., Goloubeva, O., Reddick, W. E., Glass, J. O., Gajjar, A., Kun, L., . . . Mulhern, R. K. (2001). Patterns of intellectual development among survivors of pediatric medulloblastoma: a longitudinal analysis. *Journal of Clinical Oncology, 19*(8), 2302-2308.

Platteaux, N., Dirix, P., Dejaeger, E., & Nuyts, S. (2010). Dysphagia in head and neck cancer patients treated with chemoradiotherapy. *Dysphagia, 25*, 139-152.

Platteaux, N., Dirix, P., Hermans, R., & Nuyts, S. (2010). Brachial plexopathy after chemoradiotherapy for head and neck squamous cell carcinoma. *Strahlentherapie und Onkologie, 186*(9), 517.

Radiation, U. N. S. C. o. t. E. o. A. (1988). Report to the General Assembly, with annexes. United Nations, New York.

Reddick, W. E., White, H. A., Glass, J. O., Wheeler, G. C., Thompson, S. J., Gajjar, A., Mulhern, R. K. (2003). Developmental model relating white matter volume to neurocognitive deficits in pediatric brain tumor survivors. *Cancer: Interdisciplinary International Journal of the American Cancer Society, 97*(10), 2512-2519.

Ron, E., Modan, B., Boice Jr, J. D., Alfandary, E., Stovall, M., Chetrit, A., & Katz, L. (1988). Tumors of the brain and nervous system after radiotherapy in childhood. *New England Journal of Medicine, 319*(16), 1033-1039.

Sakellariou, V. I., Badilas, N. K., Stavropoulos, N. A., Mazis, G., Kotoulas, H. K., Kyriakopoulos, S.,Sofianos, I. P. (2014). Treatment options for brachial plexus injuries. *International Scholarly Research Notices, 2014*(1), 314137.

Saremi, F., Helmy, M., Farzin, S., Zee, C. S., & Go, J. L. (2005). MRI of cranial nerve enhancement. *American Journal of Roentgenology, 185*(6), 1487-1497.

Schull, W. J., & Otake, M. (1999). Cognitive function and prenatal exposure to ionizing radiation. *Teratology, 59*(4), 222-226.

Shanley, D. (1995). Mineralizing microangiopathy: CT and MRI. *Neuroradiology, 37*, 331-333.

Sheth, S., Mukherjea, D., Rybak, L. P., & Ramkumar, V. (2017). Mechanisms of cisplatin-induced ototoxicity and otoprotection. *Frontiers in cellular neuroscience, 11*, 338.

Shorter, P., Harden, F., Owen, R., Burmeister, B., & Foote, M. (2014). Sensorineural hearing loss after treatment for head and neck cancer: a review of the literature. *Journal of Medical Imaging and Radiation Sciences, 45*(3), 316-322.

Smith, E. R., & Scott, R. M. (2010). Moyamoya: epidemiology, presentation, and diagnosis. *Neurosurgery Clinics, 21*(3), 543-551.

Spear, S. A., & Schwartz, S. R. (2011). Intratympanic steroids for sudden sensorineural hearing loss: a systematic review. *Otolaryngology--Head and Neck Surgery, 145*(4), 534-543.

Stafford, S. L., Pollock, B. E., Leavitt, J. A., Foote, R. L., Brown, P. D., Link, M. J., Schomberg, P. J. (2003). A study on the radiation tolerance of the optic nerves and chiasm after stereotactic radiosurgery. *International Journal of Radiation Oncology* Biology* Physics, 55*(5), 1177-1181.

Strojan, P., Hutcheson, K. A., Eisbruch, A., Beitler, J. J., Langendijk, J. A., Lee, A. W., Rinaldo, A. (2017). Treatment of late sequelae after radiotherapy for head and neck cancer. *Cancer treatment reviews, 59*, 79-92.

Sugiyama, H., Misumi, M., Sakata, R., Brenner, A. V., Utada, M., & Ozasa, K. (2021). Mortality among individuals exposed to atomic bomb radiation in utero: 1950–2012. *European journal of epidemiology, 36*(4), 415-428.

Taylor, A. J., Little, M. P., Winter, D. L., Sugden, E., Ellison, D. W., Stiller, C. A. Reulen, R. C. (2010). Population-based risks of CNS tumors in survivors of childhood cancer: the British Childhood Cancer Survivor Study. *Journal of Clinical Oncology, 28*(36), 5287.

Teo, P. M. L., Leung, S. F., Chan, A. T. C., Leung, T. W. T., Choi, P. H. K., Kwan, W. H., Johnson, P. J. (2000). Final report of a randomized trial on altered-fractionated radiotherapy in nasopharyngeal carcinoma prematurely terminated by significant increase in neurologic complications. *International Journal of Radiation Oncology* Biology* Physics, 48*(5), 1311-1322.

Tofilon, P. J., & Fike, J. R. (2000). The radioresponse of the central nervous system: a dynamic process. *Radiation research, 153*(4), 357-370.

Ullrich, N., Robertson, R., Kinnamon, D. D., Scott, R., Kieran, M., Turner, C., Tarbell, N. (2007). Moyamoya following cranial irradiation for primary brain tumors in children. *Neurology, 68*(12), 932-938.

Valk, P. E., & Dillon, W. P. (1991). Radiation injury of the brain. *American Journal of Neuroradiology, 12*(1), 45-62.

Vaphiades, M. S., Spencer, S. A., Riley, K., Francis, C., Deitz, L., & Kline, L. B. (2011). Radiation-induced ocular motor cranial nerve palsies in patients with pituitary tumor. *Journal of Neuro-Ophthalmology, 31*(3), 210-213.

Villari, C. R., & Courey, M. S. (2015). Management of dysphonia after radiation therapy. *Otolaryngologic Clinics of North America, 48*(4), 601-609.

Wilke, C., Grosshans, D., Duman, J., Brown, P., & Li, J. (2018). Radiation-induced cognitive toxicity: pathophysiology and interventions to reduce toxicity in adults. *Neuro-oncology, 20*(5), 597-607.

Wong, C., Fehlings, M., & Sahgal, A. (2015). Pathobiology of radiation myelopathy and strategies to mitigate injury. *Spinal Cord, 53*(8), 574-580.

Yamanaka, R., & Hayano, A. (2017). Radiation-induced malignant peripheral nerve sheath tumors: a systematic review. *World neurosurgery, 105*, 961-970. e968.

Young, D. F., Posner, J. B., Chu, F., & Nisce, L. (1974). Rapid-course radiation therapy of cerebral metastases: results and complications. *Cancer, 34*(4), 1069-1076.

Zuur, C. L., Simis, Y. J., Lansdaal, P. E., Hart, A. A., Schornagel, J. H., Dreschler, W. A., Balm, A. J. (2007). Ototoxicity in a randomized phase III trial of intra-arterial compared with intravenous cisplatin chemoradiation in patients with locally advanced head and neck cancer. *Journal of Clinical Oncology, 25*(24), 3759-3765.

Chapter 9

Infection Risks and Management in Radiation Accidents

Güliz Uyar Güleç*

Infectious Diseases and Clinical Microbiology Department,
Aydın Adnan Menderes University, Aydın, Türkiye

Abstract

Radiation accidents pose significant risks to human health and often have both immediate and long-term consequences. In addition to direct exposure to radiation, these accidents can lead to secondary infections, which exacerbate the impact on individuals and communities. The destructive effects of radiation on the immune system and body barriers are the main factors predisposing to local or systemic infections due to endogenous or exogenous microorganisms. Countries should be prepared for radiation emergencies, especially those affecting large numbers of people, and infectious disease specialists in the multidisciplinary response team should apply the most up-to-date and appropriate prophylaxis and treatment in infection management. This chapter examines the complexity of radiation accidents, their potential impact on the risk of infection, and mitigation and management strategies.

Keywords: acute radiation syndrome, infection, management, radiation accidents

---

* Corresponding Author's Email: guliz.uyar@adu.edu.tr.

In: Triage in Radiation Accidents
Editors: Nural Öztürk and Sibel Karaca
ISBN: 979-8-89530-415-0

1. Introduction

Fortunately, radiation accidents are comparatively uncommon events with other types of emergency situations. However, between 1944 and 1992, over 300 radiation accidents occurred worldwide, and over 2000 people were exposed to radiation-related adverse health effects (Dörr & Meineke, 2006). Although the number of radiation accidents has decreased over the years, the seriousness of the accidents cannot be ignored as they affect a wide range of people (Dörr & Meineke, 2006). Despite the rarity of such accidents, the potential outcomes of radiation emergencies can be very significant, including acute and chronic health impairments, psychological reactions, and notable direct and indirect economic losses. Each of these consequences requires special considerations and medical countermeasures (Maciulewicz et al., 2024).

It is of utmost importance to evaluate the patient's symptoms in great detail, as the effects of radiation exposure may be confused with those of infection, and the effects of radiation may also predispose individuals to infection. In this section, the infection risks caused by radiation accidents in acute and chronic periods and the precautions to be taken will be discussed.

2. Types of Radiation and Their Effects

As discussed in detail in previous chapters, radiation is defined as the propagation of energy in the form of electromagnetic waves or moving subatomic particles. Radioactivity occurs when atoms spontaneously disintegrate, and ionising radiation forms when excess energy is emitted into the environment (WHO, 2023). Radiation can be divided into ionising and non-ionising.

- *Ionising Radiation:* X-rays, gamma rays, and particles such as alpha and beta radiation can penetrate the body, disrupting cellular structures and DNA, leading to acute radiation syndrome (ARS) and long-term health effects such as cancer. How the effect occurs depends on the dose size and dose rate of the ionising radiation (Puckett, Al-Naser, and Nappe, 2023).
- *Non-Ionising Radiation:* While less harmful in terms of acute effects, prolonged exposure to non-ionising radiation, such as

ultraviolet (UV) and microwave radiation, can increase the risk of skin cancer and other health issues (Omer, 2021).

Since the beginning of time, all living creatures have been, and are still being, exposed to natural and man-made sources of radiation. Radiation is mainly used in various settings including industrial, medical, and military areas (USNRC, 2013).

It is a fact that radiation accidents and nuclear accidents can occur during industrial use, the use of radiation sources for medical purposes, activities in nuclear facilities, and the transportation of radioactive materials. The International Atomic Energy Agency defines an accident as a situation that has significant consequences for people, the environment or the facility. It can also occur in the form of a war or a possible nuclear terrorist attack. Recent tragic events include the Chornobyl disaster in 1986 and the Fukushima nuclear power plant accident in March 2011. In both cases, a reactor core was damaged, resulting in the release of large amounts of radiation (Cerezo, 2011).

Depending on the cause of the radiation accident, the affected population and the complications that may develop will vary (Pandey et al., 2010).

It is important to be aware that there are several ways in which radiation can cause damage. This includes external irradiation, external contamination with radioactive materials, and internal contamination by absorption through the skin, inhalation, or ingestion (López & Martín, 2011). A systematic review of reported radiation overexposure accidents worldwide between 1980 and 2013 found 634 reported radiation accidents in which 2390 people were overexposed and 190 died due to overexposure. This study found that the majority of reported radiation accidents occurred in the medical sector, with the use of radiation therapy (32%) and fluoroscopy (31%) being the main causes (Coeytaux et al., 2015).

3. Consequences of Radiation Accidents

The acute and chronic effects of radiation overexposure can be variable. In the acute period, not only the effects of radiation but also the effects of heat and trauma can be seen in casualties and their appropriate and rapid management is of vital importance (Pandey et al., 2010). Rapidly dividing human cells are the most sensitive to the harmful and lethal effects of ionising radiation (Donnelly et al., 2010).

3.1. Radyobiologic Effects

Ionising radiation causes damage to molecules through energy transmission. Ionising radiation passing through biological systems generates reactive free radicals such as free ion radicals, which further interact with critical biomolecules like DNA, proteins, lipids and membranes. This interaction subsequently results in abnormalities in cell function or cell death (Akita, 2014; Pandey et al., 2010). If lipids and proteins are extensively destroyed, the cellular DNA becomes a prime candidate for the biological and at last lethal effects of ionising free radicals. Most damage caused by ionising radiation can be repaired, and enzymatic pathways remain intact; however, radiobiological death can occur when cells lose their ability to divide and proliferate (Akita, 2014).

Reactive oxygen species (ROS), formed when water molecules are hydrolyzed by radiation, are responsible for most radiation damage. Subsequent damage is therefore chemically mediated. It is estimated that the number of ROS produced in a single human cell per day as a result of normal metabolism is approximately 1 billion. However, an effective detoxification system including antioxidants, catalases, and superoxide dismutase limits the number of DNA modifications to approximately one million. DNA alterations can be single-strand breaks, base changes, or double-strand breaks (DSBs) (Moorthy, 2021). DSBs are the primary lethal lesion (Kiang, Garrison, et al., 2010).

Furthermore, if a cell is damaged by a harmful agent such as ionising radiation, it can initiate the apoptosis cascade, also called programmed cell death. The genome is activated to enable the production of the enzyme caspase, which results in cell death (Christensen, Iddins and Sugarman, 2014).

3.2. Acute Radiation Syndrome

Acute radiation syndrome (ARS) is a broad term used to describe a short-term exposure of the whole body or specific body parts to high doses of radiation that can lead to death within hours or up to several months. It is also called acute radiation sickness. It is characterised by signs and symptoms that reflect severe damage to single or multiple organ systems (Acosta and Warrington, 2023; Macià i Garau et al., 2011).

ARS presents in four phases: the prodromal phase (0-2 days after exposure), the latent phase, the manifest disease phase and the final phase (Acosta & Warrington, 2023; López & Martín, 2011). In the prodromal phase, symptoms such as loss of appetite, nausea, vomiting, diarrhea, weakness, fatigue, fever, tachycardia and headache may be present. The latent phase is similar to the incubation period. During this phase, the person may look and feel relatively healthy externally, although internally affected by radiation. Although the subjective symptoms subside, there are still changes in places such as the blood-forming organs that will lead to the next stage. The manifest disease stage is the stage when signs and symptoms associated with radiation damage appear. Infections are particularly common at this stage. Other signs such as hair loss, fever, bleeding, severe diarrhea, weakness, disorientation and cardiovascular collapse may occur depending on the area of the body affected. Convulsions may occur at extremely high doses. The severity and onset time of signs and symptoms depend on the dose of radiation received and the onset time shortens as the dose increases. Accordingly, recovery or death occurs in the final phase (López and Martín, 2011; Macià i Garau et al., 2011). The clinical components of ARS include hematopoietic, gastrointestinal, cerebrovascular, and cutaneous sub-syndromes.

3.2.1. Hematopoietic Syndrome

Bone marrow contains multipotent primitive cells named hematopoietic stem cells (HSCs) that can develop into all blood cell types. HSCs are the source of myeloid and lymphoid lineages. All blood cells except lymphocytes are myeloid lineage cells. They include monocytes, macrophages, neutrophils, eosinophils, basophils, platelets and erythrocytes. Lymphoid cells include T cells, B cells and natural killer (NK) cells (Lee and Hong, 2020).

The hematopoietic syndrome occurs at whole-body doses of 2-3 Gray (Gy) or higher. Radiation exposure not only induces apoptosis but also alters the recirculation features of lymphocytes (López and Martín, 2011). As the absorbed dose increases, more and more hematopoietic stem cells and precursor cells will be killed and there will be no cells left to enter the post-mitotic compartment. Mature circulating cells are not affected by radiation and die at the end of their life span. The onset of signs and symptoms will depend on the balance between the rate of loss and the rate of production of circulating cells. As a result, different degrees of pancytopenia develop, with susceptibility to infection due to leukopenia and bleeding due to thrombocytopenia. Without regeneration, death usually occurs due to infection and/or hemorrhage at doses of 4.5-6 Gy without supportive care (Macià i Garau et al., 2011).

Lymphopenia can be seen within the first 6–24 hours after a moderate or high dose exposure and occurs before the onset of other cytopenias. The lymphocyte depletion point can be predicted using the lymphocyte depletion kinetics model. However, since radiation also impairs the recirculation properties of lymphocytes, their decline cannot be taken as an indicator of the extent of stem cell damage (Macià i Garau et al., 2011).

Recovery may result from the proliferation and differentiation of a non-cycling radio-resistant subpopulation of stem cells that have been preserved. These cells may play an important role in the recovery of hematopoiesis several weeks after exposure to low to moderate radiation (Harrold et al., 2020).

3.2.2. Gastrointestinal Syndrome

The gastrointestinal epithelium, which divides rapidly to shed villus cells every 2-4 days, is very sensitive to radiation (Hollingsworth et al., 2021). So, gastrointestinal syndrome occurs within 5 days after exposure. The severity of symptoms increases with increasing dose exposure. At doses <1.5 Gy patients experience nausea and vomiting initially. At doses of 5-12 Gy, abdominal cramps, diarrhea and gastrointestinal bleeding symptoms are added due to disruption of the mucosal barrier. Anemia and disturbances in fluid-electrolyte balance develop accordingly. Vomiting, severe diarrhea and high fever constitute the manifest disease. Malnutrition resulting from malabsorption and bacterial translocation, which can cause infections, leads to increased morbidity and mortality. In the last stage, severe complications may include ulceration and necrosis of the intestinal wall resulting in stenosis and perforation (Waselenko et al., 2004).

Radiation-induced cell death leads to loss of gastrointestinal epithelial integrity and function, causing both inflammation and luminal contents and microbiota to cross the epithelial barrier (Hollingsworth et al., 2021). The gastrointestinal tract, normally colonised by aerobic and anaerobic bacteria, is the source of most post-irradiation infections (Brook et al., 2004). Changes in the gut microbiota of cancer patients exposed to radiation therapy have been detected, such as an overgrowth of Gram-negative pathogenic bacilli and a decrease in protective *Bifidobacterium* and *Lactobacillus* species. These changes exacerbate radiation enteritis, reduce the absorptive surface of intestinal epithelial cells, weaken intestinal epithelial barrier function and promote inflammatory factor expression, leading to persistent mucositis, diarrhea and bacteremia (Obrador et al., 2022). In addition to these, sepsis may develop in patients with the contribution of immunosuppression that develops with the effects of hematopoietic syndrome.

Sepsis is a life-threatening organ dysfunction caused by a dysregulated host response to infection. Septic shock is a cellular and metabolic abnormality characterised by changes at both macrocirculatory and microcirculatory levels. Urgent diagnosis and treatment is required because of its high mortality rate (Gavelli et al., 2021).

Probiotics block the binding of intestinal pathogens to the host mucosa through nutrient competition, thus preventing bacterial translocation (Obrador et al., 2022). Intestinal decontamination, when used alongside antibiotics, may help reduce the population of pathogenic bacteria that translocate into the circulation and cause infection. (Christensen, Iddins and Sugarman, 2014). It is advisable to avoid the use of antacids and H2 blockers, as gastric acidity may prevent bacteria from colonising and invading the gastric mucosa, thereby reducing the incidence of nosocomial pneumonia due to aspiration (Brook et al., 2004).

3.2.3. Cerebrovascular Syndrome

The cerebrovascular syndrome, also called neurovascular syndrome, occurs at moderate or high doses (>10 Gy) due to limited cell regeneration potential and high radiosensitivity of regulatory mechanisms in cellular and subcellular components (Fliedner and Friesecke, 2001).

The prodromal phase is characterised by confusion, disorientation and fatigue and may be accompanied by dizziness and seizures. Physical examination may reveal papilledema, ataxia and abnormal deep tendon and corneal reflexes (Waselenko et al., 2004). A latent period with a marked improvement for only a few hours is followed by severe manifest disease. It is possible for watery diarrhea, respiratory distress, hyperpyrexia and cardiovascular shock to occur. As this picture may resemble the clinical course of acute sepsis and septic shock, it should be carefully evaluated for differential diagnosis (Donnelly et al., 2010; Waselenko et al., 2004).

3.2.4. Cutaneous Syndrome

Significant levels of ionising radiation can affect the skin and its appendages in two ways:

a) Exposure of the skin to whole or partial body radiation penetrating deep into the tissues (e.g., gamma, neutron),

b) Exposure of very large areas of skin, known as 'beta burns', which usually do not penetrate deep enough into the tissues to cause the other three sub-syndromes of ARS. When radiation injury from therapeutic or unintentional exposure is more localised or if the radiation dose is insufficient to penetrate the deeper organs cutaneous radiation injury occurs also known as radiodermatitis (REMM, 2024).

Clinical manifestations of radiation damage may vary depending on the dose and duration of exposure and may alter over time. The findings may include erythema, epilation, dry or moist desquamation, edema, hyperpigmentation, vascular damage, ulceration, atrophy, telangiectasia, fibrosis or necrosis (Iddins et al., 2022).

3.3. Radiation Combined Injury

Radiation-combined injury (RCI) refers to an injury that occurs when exposure to ionising radiation is combined with physical trauma. This may include damage to the skin, chemical trauma, infection and/or physiological effects (Wang et al., 2022).

RCI is more commonly seen in large-scale events such as nuclear explosions, nuclear pile accidents, nuclear terrorism, accidents involving depleted uranium weapons and clinical radiotherapy settings (Zou et al., 2008). Historically, combined injuries with simultaneous thermal burns have been observed in 60-70% of victims in Hiroshima and Nagasaki in Japan and 10% of 237 victims of radiation exposure in the Chornobyl reactor meltdown (Kiang et al., 2010).

The main features of RCI include shock, which occurs earlier and is more severe than with a simple injury; dramatic suppression of hematopoiesis and the immune system; extensive and severe gastrointestinal damage; and delayed wound healing (Zou et al., 2008).

Consequences of combined injury including acute myelosuppression and immunosuppression, increased risk of infection, fluid imbalance, macro/microcirculatory failure, massive cellular damage, and impairment of vital organ function can lead to multi-organ dysfunction and multi-organ failure. Similar to radiation exposure alone, they are the most common causes of death (Kiang et al., 2010).

The physiology of RCI is complex and not well characterised, but the observed high lethality seems likely to be due to a "cytokine storm"-like systemic inflammatory response syndrome. Many considerations must also be taken into account regarding medical management, such as the treatment of multiple injuries, critical decisions about the timing of surgery, and managing the overall higher risk of sepsis. Optimal treatments for RCI are likely to be broad-spectrum and tested on RCI models (Dicarlo et al., 2010).

3.4. Late Effects

Postponed effects of acute radiation exposure include late organ effects such as thyroid dysfunction, vascular changes, atrophy, fibrosis, cataracts and infertility. As the dose of radiation increases, the likelihood of developing cancer increases, but the cancer itself does not get worse. This is a stochastic effect (Christensen et al., 2014). Of course, in addition to physical effects, psychological effects such as anxiety disorders and post-traumatic stress disorder cannot be ignored (Kamiya et al., 2015).

3.4.1. Late Alterations in the Immune System

The deleterious effects of radiation on immune systems in the acute phase are well known. Although the survivor's immune system is restored as the hematopoietic system recovers, significant residual damage persists. Among the major immunologic changes noted are impaired T-cell function, decreased helper T-cell populations and increased levels of inflammatory cytokines in the blood. The results indicate that radiation exposure impairs the primary mechanisms responsible for maintaining T-cell homeostasis and regulating the balance between cell renewal and cell death between naive and memory T cells. Such impaired T-cell homeostasis can lead to accelerated immunological aging, chronic inflammation and major gerontological diseases such as cardiovascular diseases (Kusunoki and Hayashi, 2008).

3.4.2. Malignancy

Accidents throughout history have proven that ionising radiation is a human carcinogen. However, children have a higher risk of developing cancer after exposure to radiation than adults (Kutanzi et al., 2016). Leukemia was the first radiation-induced cancer reported in atomic bomb survivors and was prevalent predominantly in children. Surprisingly, there was no increase in childhood leukemia after the Chornobyl accident, suggesting that internal exposure to radioactive iodine is important for cancer induction compared to external exposure (Suzuki et al., 2015).

Elevated risk of solid cancers was reported about 10 years after the bombings. The results of epidemiological studies on populations living in heavily contaminated areas around Chornobyl demonstrated a significant increase in the incidence of thyroid cancer in individuals who received high thyroid doses (>1 Gy) during childhood, with the earliest observed cases occurring several years after the accident. Organs and tissues at significantly increased risk of radiation-induced cancer include the thyroid, bladder, female breast, ovary, lung, brain, colon, esophagus, stomach, liver and skin (except melanoma) (Kamiya et al., 2015).

4. Conditions Predisposing to Infection in Radiation Exposure

Irradiated victims may experience a multitude of infectious complications. For example, pulmonary infections resulting from a variety of bacterial and viral agents, infections caused by contamination of the water supply or food chain, and opportunistic skin infections resulting from burns or wounds (Dicarlo et al., 2010).

As will be discussed in detail in this section, there is a quantitative relationship between the degree of neutropenia and an increased risk of infection. Other contributory factors include a reduction in the bactericidal capacity of macrophages and granulocytes, alterations in the endogenous microbial flora, the acquisition of potentially pathogenic bacteria, changes in the defence barrier in mucous membranes and skin, and a deterioration in general health status (Brook et al., 2004).

4.1. Neutropenia

Neutrophils play a critical role in the innate immune system. They respond to invading pathogens by directly attacking bacterial or fungal cells or by releasing cytokines to mobilise inflammatory responses at the infection site (Zimmer & Freifeld, 2019). Neutropenia is defined as a decrease in the absolute number of neutrophils in the bloodstream, including fragmented cells and bands. Normal values for total WBC and absolute neutrophil count (ANC) change progressively from childhood to adolescence. Given that normal neutrophil counts vary by age and ethnicity, the lower limit of ANC is considered to be 1000/L in white children aged 2-12 months and 1500/L in those older than 12 months (Boxer, 2012).

Neutropenia can be clinically classified as;

- mild neutropenia with an ANC of 1000-1500/μL,
- moderate neutropenia with an ANC of 500-1000/μL or
- severe neutropenia with an ANC of less than 500/μL.

Some authors also define ANC counts below 200/μL as very severe neutropenia (Solomou et al., 2021). Classifying the severity of neutropenia according to ANC is useful to identify patients at high risk of potentially life-threatening infections. Patients become increasingly at risk of serious infections as neutropenia worsens, with patients with an ANC of <500 being at the highest risk (Gibson and Berliner, 2014). Bone marrow disorders affecting neutrophil production and release are associated with a much higher risk of bacterial and fungal infections than peripheral neutropenia with normal bone marrow morphology (Boxer, 2012).

4.1.1. Febrile Neutropenia

Febrile neutropenia is defined by the Infectious Diseases Society of America as oral fever ≥38.3°C (101.0°F) or persistent fever ≥38.0°C (100.4°F) for 1 hour and ANC <500/ μL or ANC expected to fall below 500/ μL within 48 hours (Freifeld et al., 2011).

It should be noted that this guideline has been essentially developed for the antimicrobial treatment of neutropenic fever episodes that arise after chemotherapy in patients with cancer. However, it should be taken into consideration that there are some differences between infections that occur after chemotherapy and those that occur after radiation accidents.

One of the distinguishing points is that the identification rate of infectious organisms in febrile episodes after chemotherapy reached only 20-30%, whereas after accidental irradiation almost all febrile episodes were documented (Dainiak 2018; Freifeld et al. 2011). Possible reasons for this difference in identification rate may be the disruption of the protective barrier function of the skin and gastrointestinal tract due to extensive radiation damage to these organs, and the often prolonged immunosuppression (typically several weeks to months) after irradiation at high radiation doses (>5 Gy). Absolute neutropenia after chemotherapy has a relatively short duration of days to weeks. However, in both conditions the sites of infection are similar, mainly skin, lung, and gastrointestinal tract (Dainiak, 2018). Possible pathogens detected in febrile neutropenic episodes are shown in Table 1. The causative pathogen may vary according to the possible focus of infection. The presumptive pathogen and local susceptibility patterns should be considered when planning prophylactic and empirical treatment (Freifeld et al., 2011; Solomou et al., 2021).

Table 1. The most commonly isolated pathogens in febrile neutropenic episodes

Gram positive organisms	Gram negative organisms
Coagulase negative staphylococci	*Escherichia coli*
Staphylococcus aureus, including methicillin-resistant strains	*Klebsiella* species
Enterococcus faecalis/faecium, including vancomycin-resistant strains	*Enterobacter* species
Viridans group streptococci	*Pseudomonas aeruginosa*
Streptococcus pneumoniae	*Acinetobacter* species
Streptococcus pyogenes	*Stenotrophomonas maltophilia*
Corynebacterium spp.	Other enteric bacteries

4.2. Superficial and Deep Effects on the Skin

The skin consists of the epidermis, dermis, and appendages including sebaceous, sweat, apocrine glands, and hair follicles (Iddins et al., 2014). As the largest organ system of the body, the skin is essential in maintaining the physiological homeostasis of the human body. Intact and healthy skin plays crucial roles in protecting against mechanical stresses and infections, fluid imbalance and thermal dysregulation. It also provides flexibility or stabilisation in some parts of the body, such as the joints (Sorg et al., 2017). Local radiation injury to the skin can occur in disciplines such as radiology, nuclear medicine, interventional radiology and interventional cardiology due to fluoroscopy, nuclear medicine and computed tomography. Many other sources of ionising radiation are industrial or commercial sources, nuclear accidents, and, potentially, terrorist events (Iddins et al., 2022). A large-scale radiological incident could affect large numbers of individuals with various degrees of ionising radiation injuries to the skin, thermal burns and traumatic injuries. Injury to the skin caused by ionising radiation may not be limited to the skin but may also affect deeper soft tissues and organs (Iddins et al., 2022). In combined radiation injury, both radiation and full-thickness skin wounding result in comparable physiological responses, including fluid loss, metabolic changes, and shock, accompanied by alterations in hormone concentration. Additionally, it results in injury to areas such as the bone marrow, gastrointestinal tract and skin, which impairs wound repair, hinders tissue healing, suppresses the immune system, increases susceptibility to microbial infections, sepsis and ultimately, death (Wang et al., 2022).

In a mice model, combined injury was shown to result in early detection of bacteria in the blood of the heart and liver. Systemic infection with *Enterococcus* sp., *Staphylococcus* sp., *Bacillus* sp. and *Lactobacillus* sp. was detected in mice with combined injury. Radiation alone resulted in delayed detection of bacteria, whereas only a transient bacteremia occurred after injury alone. The results of this study support that increased iNOS, cytokines and bacterial infection triggered by combined injury may contribute to mortality (Kiang et al., 2010).

4.3. Pulmonary Changes

There are more than 40 different types of cells in the lung and these cells alone are relatively radiation resistant. However, since the regenerative capacity of the lung is low, it is considered intolerant to high radiation doses. The lung alveoli are lined with two types of alveolar epithelial cells also known as pneumocytes: type I and type II alveolar epithelial cells. Type I pneumocytes are squamous epithelial cells that cover 90–95% of the alveolus. These flat cells are responsible for the exchange of oxygen and carbon dioxide, as well as the maintenance of fluid homeostasis. Type II cells are typical cuboidal granular cells, occupying only 7% of the total alveolar surface. They have an important role in immune regulation. They synthesize molecules such as complement system products, cytokines, lysozymes, and surfactant proteins A and D (C-type lectins). They act as stem cells for type I cells and provide epithelial repair (Lierova et al., 2018).

Exposure to ionising radiation damages the type I pneumocyte layer, weakening its barrier function. Alveolar epithelial cells that undergo apoptosis after irradiation release pro-inflammatory cytokines, which attract macrophages and strengthen the inflammatory cascade (Guo et al., 2024). As a result of these changes, radiation-induced lung injury (RILI) develops. The course of RILI usually divides into early pulmonary toxicity and late pulmonary toxicity which occurs months to years after treatment, including pulmonary tissue fibrosis, necrosis, and atrophy (Chen et al., 2023).

Radiation pneumonitis develops within days to weeks as an acute reaction. Patients often exhibit nonspecific symptoms such as fever, cough (dry or productive), shortness of breath, and chest pain. Radiation-induced pulmonary fibrosis typically becomes radiologically visible approximately six months after exposure. Furthermore, individuals who suffer from radiation pneumonitis are at an increased risk of developing radiation-associated fibrosis (Talapko et al., 2024).

Immune system weakness and alveolar epithelial damage due to radiation exposure predispose individuals to respiratory infections. These infections can further compromise respiratory system function and cause additional complications.

5. Prophylaxis and Treatment Approaches for Infections

After radiation exposure, there is increased susceptibility to infections due to disruption of cutaneous and mucosal barriers, neutropenia, and immunosuppression. Management of acute radiation syndrome and its effects on the development of infection will be presented from an infectious disease perspective. Aseptic protocols and infection control measures must be strictly followed due to the patient's suppressed immunity and increased susceptibility to infections.

The medical management of severe ARS in hospitalised patients should be multidisciplinary and should include critical care specialists, hematologists, radiologists, dermatologists, gastroenterologists, infectious disease specialists as well as radiation dosimetry specialists (Tanigawa, 2021).

The general approach to the treatment of infections in the irradiated patient can be considered in terms of time after exposure. An estimated timeline for possible infectious organisms and treatments can be given according to the duration of immune impairment after radiation overexposure. According to this timeline, the period from 0 to 3 days can be classified as the 'early' period, 3 to 30 days as the 'intermediate' period, and more than 30 days as the 'late' period (Dainiak, 2018). In the early post-exposure period, the patient is not yet neutropenic and medical resuscitation, fluid replacement and surgery, if necessary, should be prioritized (Donnelly et al., 2010; López & Martín, 2011). In the intermediate phase, CD4 count and serology of CMV, HSV, VZV, and EBV should be screened, and the addition of antiviral and antifungal agents should be evaluated together with the clinical status of the patient. In the late period, immune reconstitution can take place (Christensen et al., 2014; Dainiak, 2018). Precautions regarding gastrointestinal syndrome are mentioned above in the relevant section. The factors that pose a risk of infection and the schematic representation of the approach to infection are presented in Figure 1.

5.1. Management of Neutropenia and Neutropenic Fever

Practices in the management of neutropenia and neutropenic fever in irradiated patients are mostly based on experience with patients with hematologic or solid cancer. Three radiomitigators approved by the United States Food and Drug Administration (US FDA) for the hematopoietic syndrome are granulocyte colony-stimulating factor (G-CSF), PEGylated G-CSF, and granulocyte-macrophage colony-stimulating factor (GM-CSF) (Singh and Seed, 2020).

Granulocyte colony-stimulating factor (G-CSF) is a glycoprotein produced by activated macrophages, endothelial cells and fibroblasts. The recombinant DNA technology enables the production of G-CSF in large quantities in *Escherichia coli* and mammalian cells. In addition to its stimulatory effect on the proliferation of hematopoietic cells, the anti-apoptotic effect of G-CSF in radiation-induced myelosuppression is also worthy of note (Hofer et al., 2014).

PEGylated G-CSF is produced in yeast and chemically engineered via PEGylation for a slow, sustained pattern of drug release after administration. It is indicated in adults and children for the relief of acute neutropenia and to minimise the risk of infection and sepsis (Singh and Seed, 2021).

Natural or recombinant forms of GM-CSF are also indicated in adults and children. Recombinant GM-CSF can also activate mature macrophages. GM-CSF can stimulate and support multiple lineages, megakaryocytic and erythroid progenitors as well as the main myelogranulocyte/monocyte (Singh and Seed, 2021).

It is recommended that radiomitigators be initiated as promptly as possible following radiation exposure. A baseline complete blood count (CBC) and serial CBCs every three days until the absolute neutrophil count remains above 1000/mm^3 on three consecutive CBCs are recommended (Singh and Seed, 2020).

Three points should be considered in the initial evaluation of patients who meet the definition of neutropenic fever:

- Obtaining two sets of blood cultures and other appropriate cultures to identify the possible infectious agent

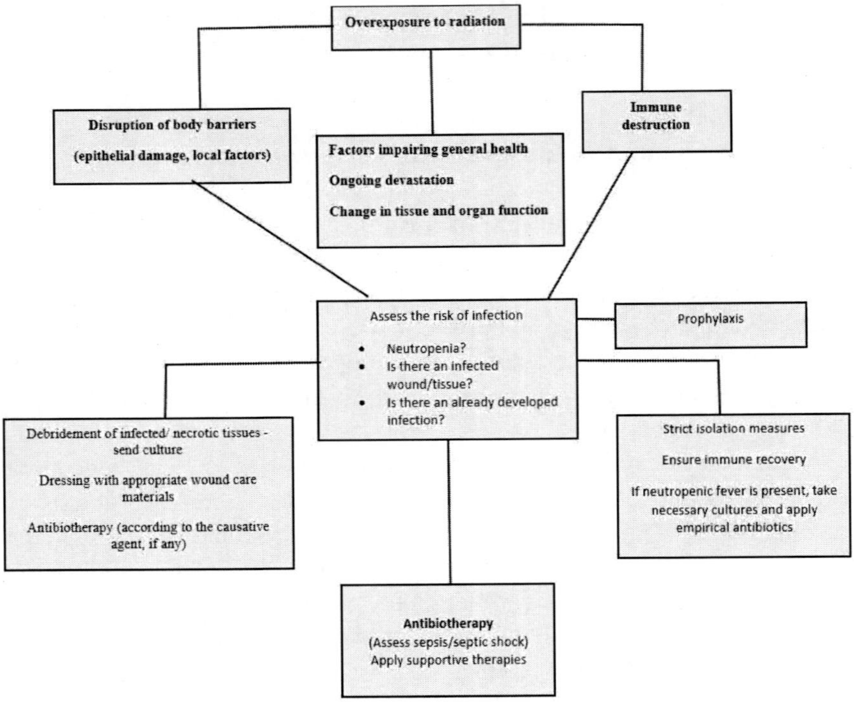

Figure 1. Infection risk factors and approach to infections.

- Imaging tests and CBC and basic biochemical examinations to determine the focus of infection
- Initiation of appropriate empirical antimicrobial medication without delay (Zimmer and Freifeld, 2019).

Patients who are expected to have absolute neutropenia lasting less than 7 days, who are hemodynamically stable, who do not have comorbidities and who are considered 'low risk' can be followed as outpatients with oral antibiotics if there are no absorption and tolerance problems (Dainiak, 2018). Empirical treatment must cover staphylococci, streptococci, pseudomonas and other Gram-negative bacteria. The combination of ciprofloxacin 500-750 mg orally every 12 hours and amoxicillin/clavulanate 500 mg orally every 8 hours is a commonly used empiric regimen (Punnapuzha et al., 2023).

Patients with the following conditions are considered part of a 'high risk' group: severe neutropenia, major uncontrolled comorbid medical conditions, including haemodynamic lability, severe mucositis, gastrointestinal manifestations, alterations of the mental status, intravenous catheter infections, pulmonary infiltrates, and underlying chronic lung disease. Additionally, patients with hepatic insufficiency, as evidenced by aminotransferase levels exceeding five times the normal value, and renal insufficiency, as indicated by creatinine clearance below 30 mL/min, are also considered at high risk (Punnapuzha et al., 2023). Hospitalisation and parenteral antibiotics are indicated for high-risk patients, for those with proven infection, and for those who cannot tolerate oral treatment (Dainiak, 2018). In the context of febrile neutropenia and in accordance with the relevant standard of care, intravenous antibiotic therapy should be commenced within an hour of initial triage. Early antibiotic administration has been shown to be associated with higher survival rates. Efforts should be made to ensure that febrile neutropenia patients receive effective antibiotic therapy as soon as possible (Rosa & Goldani, 2014). In hospitalised patients, one of the antibiotics recommended in the guidelines can be selected:

Cefepime or ceftazidime (2 g IV every 8 hours) or
Piperacillin-tazobactam (4.5 g IV every 6 to 8 hours) or

Antipseudomonal carbapenems: Imipenem-cilastatin (500 mg IV every 6 hours) or meropenem (1 to 2 g IV every 8 hours).

The addition of vancomycin should be considered for those with cellulitis or pneumonia, an aminoglycoside for those with Gram-negative bacteremia or pneumonia, and metronidazole for those with abdominal symptoms or suspected *Clostridium difficile* infection (Freifeld et al., 2011).

The length of antibiotic treatment can be determined independently of neutrophil recovery. In patients with a clinically documented infection, the administration of antibiotics may be terminated when the patient has remained afebrile for at least 72 hours without any further infectious signs or symptoms. If the patient has a microbiologically documented infection, it is recommended that the antibiotic treatment be stopped after a minimum of four days of afebrile period and seven days of antibiotic treatment. (Escrihuela-Vidal et al., 2019).

Yeasts (especially *Candida* species) and molds typically cause infections in patients with long-term neutropenia, manifested by persistent or recurrent fever. Since *Candida* species are ubiquitous colonizers of human mucosal surfaces, they can cause bloodstream infection by disruption of the mucosal barrier. The addition of empirical antifungal therapy is indicated in high-risk patients with persistent or recurrent fever after 4-7 days of broad-spectrum antibacterial therapy and with prolonged neutropenia (Walsh and Gamaletsou, 2013). Depending on local susceptibility data and the isolated agent, fluconazole, echinocandin or liposomal amphotericin B may be used as an antifungal agent in empirical treatment (Pappas et al., 2015).

In a report on a patient who was the only survivor among the three victims of the Tokai-mura criticality accident in 1999 and who was exposed to more than 2Gy of radiation in total, it was reported that the patient did not receive hematopoietic stem cell transplantation, G-CSF was administered, selective gastrointestinal tract sterilisation was performed until stool culture was negative, antifungal and antiviral drugs were administered prophylactically, and isolation measures were applied (Hirama and Akashi, 2005).

5.2. Skin and Wound Care

Radiation-induced skin injuries do not heal as quickly as burns and the closed wound has lower strength and the pain related to the injury can be much higher (Iddins et al., 2022). These wounds are more prone to bacterial infection during the healing process than conventional wounds. Imbalance in the inflammatory response, excessive accumulation of reactive oxygen species, endothelial cell damage, and impaired microvascular circulation lead to delayed wound healing. Impairment of immune function and integrity of the physical barrier creates a favourable environment for pathogen colonisation, with colonisation being the first step towards serious infection. Untreated infection impairs the process of wound healing, leading to the worsening of the wound and decreased patient quality of life (Su et al., 2022).

Most treatment regimens for patients with skin involvement are derived from traditional burn care, the side effects of radiotherapy, and the experience of those treating the small number of cutaneous radiation injuries that have occurred in historical accidents (Iddins et al., 2014).

The first treatments applied to radiation-affected skin are topical agents. Topical antihistamines are usually used for symptomatic relief of the burning and itching sensation that is common in the early phase. Silver-based topical antimicrobials can be used against infectious problems. Early assessment of antimicrobial prophylaxis and/or presumptive therapy is indicated due to the potential for compromise of the host immune system, especially in the presence of concurrent soft tissue trauma in the radiation field. Emollient moisturizers help to reduce breakage caused by dry/calloused skin. In some cases, to suppress the local immune response, topical steroids, the synergistic effect of oral or topical alpha tocopherol and/or pentoxifylline, and topical or oral non-steroidal anti-inflammatory drugs are used (Ervin et al., 2024).

Local wound care options may be traditional dressings, but it is recommended to consider hydrocolloid or hydrogel-based wound dressings. These dressings can enhance wound infection control by providing gentle adhesion, water vapour transmission with its porous structure and autolytic debridement of necrotic tissue. In the proliferative stage, hydrogels can stimulate vascular regeneration and fibroblast proliferation through the release of growth factors and degradation of bioactive substances (Shu et al., 2021).

Initial surgical debridement and wound closure methods are one of the important treatments for radiation wounds. However, the intraoperative wound appearance can be deceptive, the graft may fail to adhere and necrosis may develop due to impaired viability of the surrounding soft tissues. Better success has been reported with coverage provided by products such as biosynthetic or bioengineered epidermal or dermal skin grafts (Ervin et al., 2024). Surgery also is the main component of the treatment of chronic ulcers and fibrosis (Akita, 2014).

Conclusion

In individuals exposed to radiation, many conditions that predispose to infection may coexist. The most important of these is impaired immune system functions. In addition, the destruction of major barriers such as skin and gastrointestinal system epithelium leads to the passage of bacteria and prepares the ground for systemic infections. The presence of accompanying pathologies such as shock, heat and physical trauma facilitates the development of infection. As a result of the pathological processes that continue after exposure, changing findings may be encountered in the follow-up of the patient. For example, the development of leucopenia (neutropenia and lymphopenia) is an important risk factor for infection. However, the main difficulty in managing radiation-exposed patients is the lack of strong evidence. In the absence of randomised controlled trials of medical countermeasures in people with ARS, specific recommendations are largely based on the results of experimental animal studies and published guidelines for the treatment of unirradiated individuals.

It is essential to prevent life-threatening infections in individuals exposed to radiation at doses that cause suppression of the immune system and to manage their treatment appropriately. Firstly, the patient should be evaluated for the need for prophylaxis and necessary medications should be administered. It is essential to follow up the clinical and laboratory values of the patient, to take the necessary cultures in suspicious cases, and to start the appropriate empirical treatment immediately. Patients with radiation-induced wounds and wound infections should be treated with current therapies.

References

Acosta, R., & Warrington, S. (2023). *Radiation Syndrome.* In: StatPearls [Internet]. Treasure Island (FL): StatPearls Publishing PMID: 28722960.

Akita, S. (2014). Treatment of radiation injury. *Adv Wound Care (New Rochelle).*, *3*(1), 1–11. https://doi.org/10.1089/wound.2012.0403.

Boxer, L. A. (2012). How to approach neutropenia. *Hematology / the Education Program of the American Society of Hematology. American Society of Hematology. Education Program*, *2012*, 174–182. https://doi.org/10.1182/asheducation.v2012.1.174.3798251.

Brook, I., Elliott, T. B., Ledney, G. D., Shoemaker, M. O., & Knudson, G. B. (2004). Management of Postirradiation Infection: Lessons Learned from Animal Models. *Military Medicine*, *169*(3), 194–197. https://doi.org/10.7205/MILMED.169.3.194.

Cerezo, L. (2011). Radiation accidents and incidents. What do we know about the medical management of acute radiation syndrome? *Reports of Practical Oncology and Radiotherapy*, *16*(4), 119–122. https://doi.org/10.1016/j.rpor.2011.06.002.

Chen, Z., Wang, B., Wu, Z., Xiao, H., Yang, Y., Fan, J., Gu, Y., Chen, C., & Wu, J. (2023). The occurrence and development of radiation-induced lung injury after interstitial brachytherapy and stereotactic radiotherapy in SD rats. *Journal of Inflammation (United Kingdom)*, *20*(1), 1–11. https://doi.org/10.1186/s12950-023-00348-9.

Christensen, D. M., Iddins, C. J., Parrillo, S. J., Glassman, E. S., & Goans, R. E. (2014). Management of ionizing radiation injuries and illnesses, part 4: Acute radiation syndrome. *Journal of the American Osteopathic Association*, *114*(9), 702–711. https://doi.org/10.7556/jaoa.2014.138.

Christensen, D. M., Iddins, C. J., & Sugarman, S. L. (2014). Ionizing radiation injuries and illnesses. *Emergency Medicine Clinics of North America*, *32*(1), 245–265. https://doi.org/10.1016/j.emc.2013.10.002.

Coeytaux, K., Bey, E., Christensen, D., Glassman, E. S., Murdock, B., & Doucet, C. (2015). Reported radiation overexposure accidents worldwide, 1980-2013: A systematic review. *PLoS ONE, 10*(3), 1–26. https://doi.org/10.1371/journal.pone.0118709.

Dainiak, N. (2018). Medical management of acute radiation syndrome and associated infections in a high-casualty incident. *Journal of Radiation Research*, *59*(March), ii54–ii64. https://doi.org/10.1093/jrr/rry004.

Dicarlo, A. L., Ramakrishnan, N., & Hatchett, R. J. (2010). Radiation combined injury: Overview of NIAID research. *Health Physics*, *98*(6), 863–867. https://doi.org/10.1097/HP.0b013e3181a6ee32.

Donnelly, E. H., Nemhauser, J. B., Smith, J. M., Kazzi, Z. N., Farfán, E. B., Chang, A. S., & Naeem, S. F. (2010). Acute radiation syndrome: Assessment and management. *Southern Medical Journal*, *103*(6), 541–546. https://doi.org/10.1097/SMJ.0b013e3181ddd571.

Dörr, H. D., & Meineke, V. (2006). Appropriate radiation accident medical management: Necessity of extensive preparatory planning. *Radiation and Environmental Biophysics*, *45*(4), 237–244. https://doi.org/10.1007/s00411-006-0068-x.

Ervin, M. D., Goans, R., Diffenderfer-Stewart, K., Aloisi, B., & Iddins, C. J. (2024). Cutaneous Radiation Injuries: REAC/TS Clinical Experience. *Disaster Medicine and Public Health Preparedness*, *18*. https://doi.org/10.1017/dmp.2023.233.

Escrihuela-Vidal, F., Laporte, J., Albasanz-Puig, A., & Gudiol, C. (2019). Update on the management of febrile neutropenia in hematologic patients. *Revista Espanola de Quimioterapia : Publicacion Oficial de La Sociedad Espanola de Quimioterapia*, *32*, 55–58.

Fliedner T. M., Friesecke I., B. K. (2001). *Medical Management of Radiation Accidents: Manual on the Acute Radiation Syndrome. (METREPOL).* (originally published by Oxford: British Institute of Radiology; 2001) Compendium to the main METREPOL document.

Freifeld, A. G., Bow, E. J., Sepkowitz, K. A., Boeckh, M. J., Ito, J. I., Mullen, C. A., Raad, I. I., Rolston, K. V., Young, J. A. H., & Wingard, J. R. (2011). Clinical practice guideline for the use of antimicrobial agents in neutropenic patients with cancer: 2010 Update by the Infectious Diseases Society of America. *Clinical Infectious Diseases*, *52*(4). https://doi.org/10.1093/cid/cir073.

Gavelli, F., Castello, L. M., & Avanzi, G. C. (2021). Management of sepsis and septic shock in the emergency department. *Internal and Emergency Medicine*, *16*(6), 1649–1661. https://doi.org/10.1007/s11739-021-02735-7.

Gibson, C., & Berliner, N. (2014). How we evaluate and treat neutropenia in adults. *Blood*, *124*(8), 1251–1258. https://doi.org/10.1182/blood-2014-02-482612.

Guo, H., Yu, R., Zhang, H., & Wang, W. (2024). Cytokine, chemokine alterations and immune cell infiltration in Radiation-induced lung injury: Implications for prevention and management. *International Immunopharmacology*, *126* (November 2023), 111263. https://doi.org/10.1016/j.intimp.2023.111263.

Harrold, J., Gisleskog, P. O., Perez-Ruixo, J. J., Delor, I., Chow, A., Jacqmin, P., & Melhem, M. (2020). Prediction of Survival Benefit of Filgrastim in Adult and Pediatric Patients With Acute Radiation Syndrome. *Clinical and Translational Science*, *13*(4), 807–817. https://doi.org/10.1111/cts.12777.

Hirama, T., & Akashi, M. (2005). Multi-organ involvement in the patient who survived the Tokai-mura criticality accident. *The British Journal of Radiology, Supplement*(1), 17–20. https://doi.org/10.1259/bjr/51028846.

Hofer, M., Pospíšil, M., Komurková, D., & Hoferová, Z. (2014). Granulocyte colony-stimulating factor in the treatment of acute Radiation Syndrome: A concise review. *Molecules*, *19*(4), 4770–4778. https://doi.org/10.3390/molecules19044770.

Hollingsworth, B. A., Cassatt, D. R., DiCarlo, A. L., Rios, C. I., Satyamitra, M. M., Winters, T. A., & Taliaferro, L. P. (2021). Acute Radiation Syndrome and the Microbiome: Impact and Review. *Frontiers in Pharmacology*, *12*(May), 1–24. https://doi.org/10.3389/fphar.2021.643283.

Iddins, C. J., Christensen, D. M., Parrillo, S. J., Glassman, E. S., & Goans, R. E. (2014). Management of ionizing radiation injuries and illnesses, Part 5: Local radiation injury. *Journal of the American Osteopathic Association*, *114*(11), 840–848. https://doi.org/10.7556/jaoa.2014.170.

Iddins, C. J., Dicarlo, A. L., Ervin, M. D., Herrera-Reyes, E., & Goans, R. E. (2022). Cutaneous and local radiation injuries. *Journal of Radiological Protection*, *42*(1), 1–38. https://doi.org/10.1088/1361-6498/ac241a.

Kamiya, K., Ozasa, K., Akiba, S., Niwa, O., Kodama, K., Takamura, N., Zaharieva, E. K., Kimura, Y., & Wakeford, R. (2015). Long-term effects of radiation exposure on health. *The Lancet*, *386*(9992), 469–478. https://doi.org/10.1016/S0140-6736(15)61167-9.

Kiang, J. G., Garrison, B. R., & Gorbunov, N. V. (2010). Radiation Combined Injury: DNA Damage, Apoptosis, and Autophagy. *Adaptive Medicine*, *2*(1), 1–10. https://doi.org/
10.4247/am.2010.aba004.

Kiang, J. G., Jiao, W., Cary, L. H., Mog, S. R., Elliott, T. B., Pellmar, T. C., & Ledney, G. D. (2010). Wound trauma increases radiation-induced mortality by activation of iNOS pathway and elevation of cytokine concentrations and bacterial infection. *Radiation Research*, *173*(3), 319–332. https://doi.org/10.1667/RR1892.1.

Kusunoki, Y., & Hayashi, T. (2008). Long-lasting alterations of the immune system by ionizing radiation exposure: Implications for disease development among atomic bomb survivors. *International Journal of Radiation Biology*, *84*(1), 1–14. https://doi.org/10.1080/09553000701616106.

Kutanzi, K. R., Lumen, A., Koturbash, I., & Miousse, I. R. (2016). Pediatric Exposures to Ionizing Radiation: Carcinogenic Considerations. *International Journal of Environmental Research and Public Health*, *13*(11). https://doi.org/10.3390/ijerph13111057.

Lee, J. Y., & Hong, S. H. (2020). Hematopoieticstem cells and their roles in tissueregeneration. *International Journal of Stem Cells*, *13*(1), 1–12. https://doi.org/10.15283/ijsc19127.

Lierova, A., Jelicova, M., Nemcova, M., Proksova, M., Pejchal, J., Zarybnicka, L., & Sinkorova, Z. (2018). Cytokines and radiation-induced pulmonary injuries. *Journal of Radiation Research*, *59*(6), 709–753. https://doi.org/10.1093/jrr/rry067.

López, M., & Martín, M. (2011). Medical management of the acute radiation syndrome. *Reports of Practical Oncology and Radiotherapy*, *16*(4), 138–146. https://doi.org/10.1016/j.rpor.2011.05.001.

Macià i Garau, M., Lucas Calduch, A., & López, E. C. (2011). Radiobiology of the acute radiation syndrome. *Reports of Practical Oncology and Radiotherapy*, *16*(4), 123–130. https://doi.org/10.1016/j.rpor.2011.06.001.

Maciulewicz, T. S., Kazzi, Z., Navis, I. L., Nelsen, G. J., Cieslak, T. J., Newton, C., Lin, A., West, D. J., & Walter, F. G. (2024). Pediatric Medical Countermeasures: Antidotes and Cytokines for Radiological and Nuclear Incidents and Terrorism. *Disaster Medicine and Public Health Preparedness*, *18*, e76. https://doi.org/10.1017/dmp.2024.35.

Moorthy, S. (2021). How Safe Are Radiation Doses in Diagnostic Radiology? A Historical Perspective and Review of Current Evidence. *Indian Journal of Radiology and Imaging*, *31*(3), 653–660. https://doi.org/10.1055/s-0041-1735927.

Obrador, E., Salvador-Palmer, R., Villaescusa, J. I., Gallego, E., Pellicer, B., Estrela, J. M., & Montoro, A. (2022). Nuclear and Radiological Emergencies: Biological Effects, Countermeasures and Biodosimetry. *Antioxidants*, *11*(6), 1–51. https://doi.org/10.3390/antiox11061098.

Omer, H. (2021). Radiobiological effects and medical applications of non-ionizing radiation. *Saudi Journal of Biological Sciences*, *28*(10), 5585–5592. https://doi.org/10.1016/j.sjbs.2021.05.071.

Pandey, B. N., Kumar, A., Tiwari, P., & Mishra, K. P. (2010). Radiobiological basis in management of accidental radiation exposure. *International Journal of Radiation Biology*, *86*(8), 613–635. https://doi.org/10.3109/09553001003746059.

Pappas, P. G., Kauffman, C. A., Andes, D. R., Clancy, C. J., Marr, K. A., Ostrosky-Zeichner, L., Reboli, A. C., Schuster, M. G., Vazquez, J. A., Walsh, T. J., Zaoutis, T. E., & Sobel, J. D. (2015). Clinical Practice Guideline for the Management of Candidiasis: 2016 Update by the Infectious Diseases Society of America. *Clinical Infectious Diseases*, *62*(4), e1–e50. https://doi.org/10.1093/cid/civ933.

Punnapuzha, S., Edemobi, P., & Elmoheen, A. (2023). *Febrile Neutropenia.* In: StatPearls [Internet]. Treasure Island (FL): StatPearls Publishing. https://www.ncbi.nlm.nih.gov/books/NBK541102/.

REMM. (2024). *Cutaneous Radiation Syndrome*. Guidance on Diagnosis and Treatment for Healthcare Providers. https://remm.hhs.gov/ cutaneoussyndrome.htm#evaluation.

Rosa, R. G., & Goldani, L. Z. (2014). Cohort study of the impact of time to antibiotic administration on mortality in patients with febrile neutropenia. *Antimicrobial Agents and Chemotherapy*, *58*(7), 3799–3803. https://doi.org/10.1128/AAC.02561-14.

Shu, W., Wang, Y., Zhang, X., Li, C., Le, H., & Chang, F. (2021). Functional Hydrogel Dressings for Treatment of Burn Wounds. *Frontiers in Bioengineering and Biotechnology*, *9*(December), 1–18. https://doi.org/10.3389/fbioe.2021.788461.

Singh, V. K., Seed, T. M. (2021). Radiation countermeasures for hematopoietic acute radiation syndrome: growth factors, cytokines and beyond. *International Journal of Radiation Biology*, *97*(11), 1526–1547. https://doi.org/10.1080/09553002.2021.1969054.

Singh, V., Seed, T. (2020). Pharmacological management of ionizing radiation injuries: current and prospective agents and targeted organ systems. *Expert Opin Pharmacother*, *21*(3), 317–337. https://doi.org/10.1080/14656566.2019.1702968. Pharmacological.

Solomou, E. E., Salamaliki, C., & Lagadinou, M. (2021). How to Make the Right Diagnosis in Neutropenia. *Clinical Hematology International*, *3*(2), 41. https://doi.org/10.2991/
chi.k.210216.001.

Sorg, H., Tilkorn, D. J., Hager, S., Hauser, J., & Mirastschijski, U. (2017). Skin Wound Healing: An Update on the Current Knowledge and Concepts. *European Surgical Research*, *58*(1–2), 81–94. https://doi.org/10.1159/000454919.

Su, Y., Cui, H., Yang, C., Li, L., Xu, F., Gao, J., & Zhang, W. (2022). Hydrogels for the treatment of radiation-induced skin and mucosa damages: An up-to-date overview. *Frontiers in Materials*, *9*(October), 1–18. https://doi.org/10.3389/fmats.2022.1018815.

Suzuki, K., Mitsutake, N., Saenko, V., & Yamashita, S. (2015). Radiation signatures in childhood thyroid cancers after the Chernobyl accident: Possible roles of radiation in carcinogenesis. *Cancer Science*, *106*(2), 127–133. https://doi.org/10.1111/cas.12583.

Talapko, J., Talapko, D., Katalinić, D., Kotris, I., Erić, I., Belić, D., Vasilj Mihaljević, M., Vasilj, A., Erić, S., Flam, J., Bekić, S., Matić, S., & Škrlec, I. (2024). Health Effects of Ionizing Radiation on the Human Body. *Medicina*, *60*(4), 653. https://doi.org/10.3390/MEDICINA60040653.

Tanigawa, K. (2021). Case review of severe acute radiation syndrome from whole body exposure: concepts of radiation-induced multi-organ dysfunction and failure. *Journal of Radiation Research*, *62*, I15–I20. https://doi.org/10.1093/jrr/rraa121.

Walsh, T. J., & Gamaletsou, M. N. (2013). Treatment of fungal disease in the setting of neutropenia. *Hematology / the Education Program of the American Society of Hematology. American Society of Hematology. Education Program*, *2013*, 423–427. https://doi.org/10.1182/asheducation-2013.1.423.

Wang, L., Lin, B., Zhai, M., Cui, W., Hull, L., Zizzo, A., Li, X., Kiang, J. G., & Xiao, M. (2022). Deteriorative Effects of Radiation Injury Combined with Skin Wounding in a Mouse Model. *Toxics*, *10*(12), 1–17. https://doi.org/10.3390/toxics10120785.

Waselenko, J. K., MacVittie, T. J., Blakely, W. F., Pesik, N., Wiley, A. L., Dickerson, W. E., Tsu, H., Confer, D. L., Coleman, C. N., Seed, T., Lowry, P., Armitage, J. O., & Dainiak, N. (2004). Medical management of the acute radiation syndrome: Recommendations of the Strategic National Stockpile Radiation Working Group. *Annals of Internal Medicine*, *140*(12). https://doi.org/10.7326/0003-4819-140-12-200406150-00015.

WHO. (2023). *No Title*. Ionizing Radiation and Health Effects. https://www.who.int/news-room/fact-sheets/detail/ionizing-radiation-and-health-effects.

Zimmer, A. J., & Freifeld, A. G. (2019). Optimal management of neutropenic fever in patients with cancer. *Journal of Oncology Practice*, *15*(1), 19–24. https://doi.org/10.1200/JOP.18.00269.

Zou, Z., Sun, H., Su, Y., Cheng, T., & Luo, C. (2008). Progress in research on radiation combined injury in China. *Radiation Research*, *169*(6), 722–729. https://doi.org/10.1667/RR1284.1.

Chapter 10

Radiation Awareness of the Public and Radiation Workers

Sibel Karaca*

Department of Radiation Oncology, Akdeniz University, Antalya, Türkiye

Abstract

Awareness is a relative concept. The main emphasis often revolves around an individual's internal state, such as emotional experiences or external occurrences observed through sensory mechanisms. It is crucial to distinguish awareness from the acts of observation and perception, which involve the recognition of various sensations. Since radiation is not perceptible to the human senses (sight, hearing and touch), it poses a challenge in terms of awareness. After the groundbreaking discovery of X-rays in 1895 and the subsequent understanding of radioactivity, the existence of radiation rapidly became widespread and its ubiquity, whether through natural exposure or technological advances, underscores its importance in society. The proliferation of nuclear power plants and medical technologies has contributed to an escalation in radiation exposure for both the general population and workers. Despite the myriad benefits associated with the use of radiation, it is essential to recognise its adverse effects on living organisms. Improving knowledge and awareness of radiation hazards is essential to ensure compliance with safety protocols. The use of social media platforms and educational initiatives play a crucial role in disseminating information and raising awareness of radiation-related risks among the public and employees. This section of the article

* Corresponding Author's Email: sibeltuzlaci@gmail.com.

In: Triage in Radiation Accidents
Editors: Nural Öztürk and Sibel Karaca
ISBN: 979-8-89530-415-0

provides a comprehensive overview of radiation awareness among the public and radiation workers.

Keywords: awareness, public, radiation, radiation workers

1. Introduction

Radiation is generally defined as the transfer of energy in space or matter in the form of particles or waves. Ionising radiation, as defined by the World Health Organisation, can originate from natural background radiation or artificial sources (WHO, 2023). Human exposure is unavoidable due to the ubiquitous presence of natural background radiation. Sources of background radiation include cosmic rays, gamma rays from space, radioisotopes and radon, a radioactive gas. In addition to natural sources, man-made radiation has been used extensively since the discovery of X-rays. Due to their small wavelengths, gamma rays, X-rays and cosmic rays have sufficient energy to break chemical bonds, resulting in ionisation. This category of radiation is therefore known as ionising radiation.

People are exposed to ionising radiation through their work in many fields, such as industry, medicine, education, research, nuclear power generation and fuel production. In medicine, ionising radiation sources are used in an increasing number and variety of ways in many different diagnostic and therapeutic areas (Branner et al., 2003; Mettler and Upton, 2008). Healthcare workers (physicians, technicians, health physicists, nurses, etc.) working in areas such as nuclear medicine, radiotherapy, operating theatres, interventional radiology and angiography units, where ionising radiation is used extensively, are at high risk of being exposed to harmful effects of radiation (Cupitt et al., 2001). Workers in construction, mining, agriculture and nuclear power are also at risk. Unfortunately, not only workers but also the general public can be exposed to the harmful effects of radiation. The exposure of patients and their families to radiation increases, especially during medical diagnosis and treatment. Studies conducted in recent years show that the number of tomographic scans has increased dramatically and that low-dose radiation applications can have adverse effects on human health (Dehghani et al., 2014). Humans are not only exposed to ionising radiation, but also to non-ionising radiation. Several commonly used electronic devices that produce non-ionising radiation (television, radio, computer, mobile phone etc.) are known to have adverse effects on living organisms (Havas, 2017).

The WHO highlights the negative consequences of excessive exposure to ionising radiation (Allam et al., 2024). Unfortunately, it is not always possible to be completely protected from natural radiation. However, it may be possible to protect oneself from some sources of radiation. Protection is particularly important in medical applications or in nuclear power stations, which are artificially produced and have become more widespread in recent years. As radiation has many useful applications, it is not possible to eliminate it from human life.

Since ionising radiation poses a risk to both the public and, in particular, radiation workers, it is necessary to be knowledgeable about radiation protection and to take the risk seriously (Ramanathan and Ryan, 2015). Through awareness, people's fields of consciousness expand towards their environment and the universe. To increase radiation awareness, the public and radiation workers need comprehensive information about the dangers of radiation exposure and radiation protection.

2. What Is Radiation?

Radiation can be defined as energy moving through space or matter in the form of particles or waves. Radiation is produced in two ways. One is the radioactive decay of an unstable atom (radionuclide) and the other is the interaction of a particle with matter. Atoms are the basic structural unit of matter and the defining structure of the elements. The atomic structure consists of a nucleus of protons, neutrons and electrons orbiting the nucleus (Figure 1). The nucleus carries a positive charge; protons are positively charged, electrons are negatively charged and neutrons carry no charge. Neutrons and photons are heavier than electrons and reside in the nucleus at the centre of the atom. If the number of neutrons in the nucleus is much higher than the number of protons, the substance is unstable. Neutrons decay and emit β^- (negatrons). If there are more protons than neutrons, the protons transform and emit β^+ (positrons). In an unstable nucleus, the neutrons and protons that leave the nucleus emit gamma (γ) rays, and heavy nuclei can emit alpha (α) rays (Keith et al., 2012). But sometimes it occurs through the absorption of one of the orbital electrons by the nucleus and the subsequent emission of characteristic x-rays. This property of a radioisotope is known as radioactivity. In this way, the atom becomes a stable isotope of another element; for example, the radioisotope iodine 125 becomes the stable isotope tellurium 125 (Mehta, 2005).

The process of radioactive decay is random. It is difficult to predict when a particular atom will start to decay. However, it is possible to measure the number of isotopes in a group that decay over a period of time. This rate of decay is known as radioactivity. The half-life of a radioactive material is proportional to its number of atoms (Mehta et al., 2010).

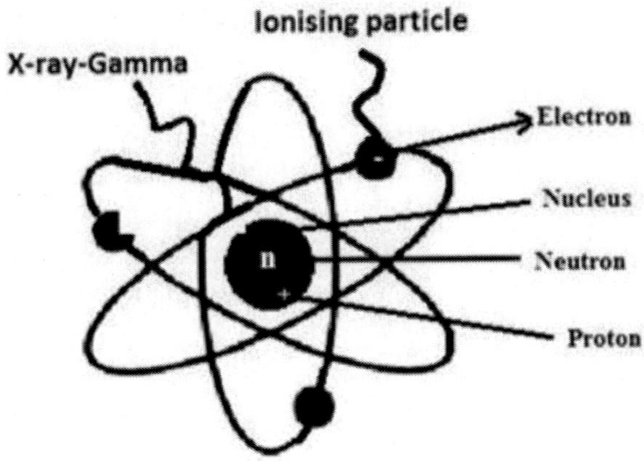

Figure 1. Structure of Atom.

Roentgen discovered radiation (he called it X-rays) at the University of Würzburg in Germany in 1985 (Glasser, 1995). X-rays are produced artificially in X-ray tubes. Electrons from a heated cathode are accelerated by a voltage of tens of thousands of volts and collide with the target anode in the tube; when the electrons are stopped as a result of this collision, the energy lost by the electrons is emitted as X-rays. This phenomenon is called the Bremstrahlung phenomenon, and the continuous spectrum formed by the resulting X-rays is called Bremstrahlung radiation. Radiation has been present on Earth since the beginning of the world. There are many types of radiation. Radio waves, microwaves, infrared radiation, visible radiation, X-rays and gamma rays are examples. Figure 2 shows the electromagnetic spectrum.

There are two kinds of radiation in nature. These are called ionising and non-ionising radiation. Ionising radiation includes radiation from both natural and man-made radioactive materials. Ionising radiation is further divided into direct ionising radiation (protons, electrons and ions) and indirect ionising radiation (γ, n, x, etc.). The average energy required to produce ion pairs in air is 34 eV. Therefore, a few tens of keV can be considered as the boundary between ionising and non-ionising radiation. Electrically charged molecules and atoms are called ions. Non-ionising radiation has less energy than ionising radiation; it does not have enough energy to produce ions (remove electrons from atoms). Examples of non-ionising radiation include visible light, infrared, radio waves, microwaves and sunlight. Ionising radiation can disrupt the electron-proton balance by removing electrons from their orbits around atoms and giving the atom a positive charge (Attix, 1988; Podgorsak, 2005).

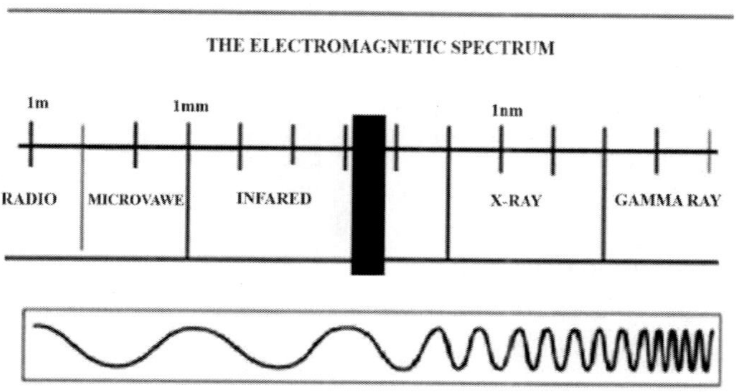

Figure 2. Electromagnetic Spectrum.

3. Radiation in Life

Since the discovery of X-rays, radiation has been used in many fields, including medicine, industry, energy, agriculture and many others. Natural sources of radiation include natural radioactive isotopes found in the Earth's crust (soil, air, water, plants and other living things) and high-energy cosmic rays that enter the Earth's atmosphere from space. The sources of radiation to which the public is exposed are shown in Figure 3.

The use of X-rays in medicine has revolutionised medicine. The use of medical radiation can be grouped under three headings: radiodiagnostics, radiotherapy and nuclear medicine. X-ray imaging (radiography, fluoroscopy, mammography and computed tomography) has become widespread in modern medicine. In every medium-sized hospital in the developed world, the number of radiotherapy units has increased, providing treatment for many cancer patients. In addition, the diagnostic and therapeutic properties of radiation are widely used in nuclear medicine. Medical radiation is the second largest source of radiation after natural radiation, accounting for 20-50% of personal radiation doses (Feng et al., 2022; Mettler et al., 2020). According to the UNSCEAR 2000 Survey Report, the total average annual dose to the world population is 2.8 mSv. Of this total dose, 14% is due to natural cosmic radiation, 18% to natural external radiation, 11% to natural internal radiation, 14% to medical radiation, 0.25% to nuclear industry, 43% to natural radon and 14% to natural cosmic radiation (UNESCO, 2000).

Radiation has many beneficial applications in human life. Examples of these applications are,

- Diagnosing (radiological imaging) and treating disease (radiotherapy and nuclear medicine),
- Protecting the health of farm animals,
- Development of water resources,
- Controlling harmful insects,
- Preserving food,
- Strengthening industrial quality control,
- Tissue sterilisation,
- Production and sterilisation of medical materials,
- Applications in the polymer industry (degradation of Teflon and cellulose, cross-linking of plastic pipes, cables and wires),
- Environmental applications,
- Industrial products manufactured using radiation technology.

Radiation in the environment can reach people in many ways. For example, radioactive material in the air can fall into food grown in agricultural areas. People can then eat these foods and therefore be exposed to this radiation or people can easily inhale radioactive substances in the air. People can be exposed to radon gas inside their homes. While walking on the road, they may be exposed to cosmic radiation (from the sun and stars) or radiation caused by the presence of naturally occurring radioactive substances such as uranium, thorium, and radium in soil, water, and rocks. All organic matter (plant and animal) also contains radioactive carbon and potassium. Additionally, background radiation dose levels from these sources vary in different parts of the world. Geological differences and the amount of radiation exposed are effective (Zakariya and Kahn, 2014). The public is constantly irradiated for diagnosis and treatment purposes in healthcare institutions or with artificial radiation through many technological products. Radiation exposure levels can be divided into three categories:

- High level
- Medium level
- Low level

Radiation exposure at a dose that causes significant damage to the body at high levels of exposure can quickly kill the exposed person. Damaged cells cannot be quickly repaired. Exposure to radiation from nuclear weapons results in high doses. In some cases, however, high doses in a controlled situation can be beneficial. Radiotherapy is the best example. High doses of radiation are doses above 1000 mSv. Intermediate-level radiation will not kill a person, but it can damage reproductive cells or other cells in the body. Permanently damaged or altered cells may continue to produce abnormal cells when they divide. Under these conditions, these cells can become cancerous. It can take many years for cancer to appear after this type of exposure. Moderate doses of radiation are in the hundreds of mSv. However, radiation-induced hereditary effects have not yet been identified in human populations exposed to radiation, although they are known to occur in other species. Low level radiation exposure is exposure to natural background radiation. It can damage reproductive cells or cause cancer. Low-level radiation doses are in the tens of mSv. The effects of low doses of radiation on the cell may not be observed for many years (Zakariya and Kahn, 2014).

Despite the many beneficial properties of radiation, it can cause biological and environmental damage if care is not taken. When ionising radiation comes into contact with the human body due to its penetrating properties, it transfers energy into the body's structure and can thus interfere with the structure of molecules. In the human body, this energy transfer can disrupt or destroy cellular functions. In addition, it can change the genetic code of cells. The cellular effects of radiation can occur in two ways. These are deterministic and stochastic effects. In deterministic effects, the severity is proportional to the dose (threshold), whereas in stochastic effects, the probability, not the severity, is proportional to the dose (Frischknecht, 2000).

Therefore, deterministic (acute) effects occur only when the radiation dose is significant, as in accidents. Stochastic effects (cancer and hereditary effects) can result from damage to a single cell.

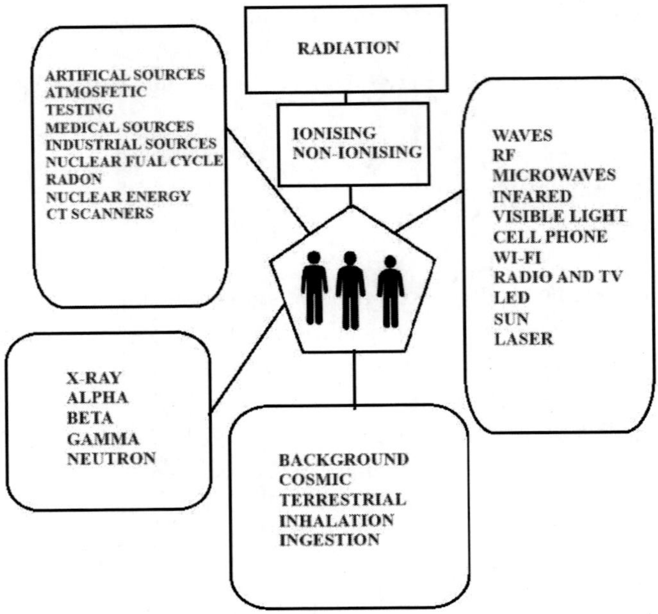

Figure 3. Types of Radiation Public Exposure.

The effects of radiation depend on many factors, including

- The type of radiation (X-ray, alpha, beta or gamma)
- Duration of exposure (regular, low, short or high t)
- Quantity
- Ratio
- Area irradiated
- Age of the person

Radiation exposure is sometimes accidental and sometimes intentional (nuclear reactor explosion, atomic bomb, radiotherapy, CT imaging, radon, etc.). Radiation exposure may be external or internal. To avoid internal exposure, the environment should be washed and thoroughly cleaned to adequately reduce dust in the work area. The environment should be well ventilated. Appropriate respiratory protection should be used. Protective clothing and equipment should be used, and work areas should be kept clean.

The three basic principles of radiation protection are justification, optimisation and dose limitation. The As Low As Reasonably Achievable

(ALARA) principle, defined in international regulations, was established to ensure that all measures are taken to reduce radiation exposure, recognising that radiation is an integral part of the diagnosis and treatment of patients. Exposure to radiation, even in small doses, increases the risk of stochastic effects, or the likelihood of developing cancer after radiation exposure (Rosen et al., 2015). Organisations such as the International Committee on Radiation Units (ICRU), the International Commission on Radiological Protection (ICRP) and the International Atomic Energy Agency (IAEA) are responsible for setting radiation protection guidelines (Feng et al., 2022). There are three principles for reducing radiation exposure: time, distance and shielding (Figure 4).

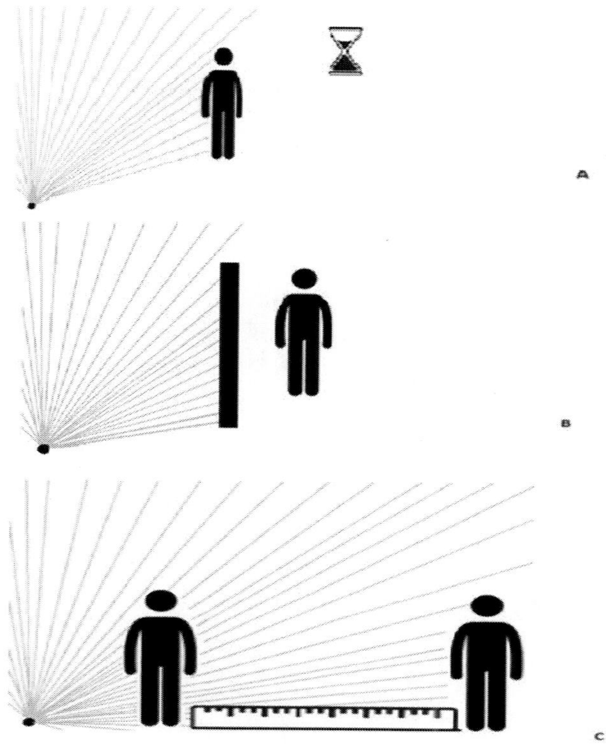

Figure 4. Basic Principles of Protection from External Radiation (A: Time, B: Shielding, C: Distance).

Radiation doses can be limited as much as possible by reducing the duration of exposure. Exposure time can be minimised in a number of ways.

Increasing the distance from the source is another way to minimise exposure. Scatter exposure levels decrease in proportion to the inverse square of the distance from the X-ray source. By doubling the distance from the source, exposure levels can be reduced by a factor of four. This simple concept can significantly reduce occupational exposure. It has been shown to reduce the effective radiation dose to personnel by more than 90% when appropriate shielding is used (López et al., 2018).

The maximum allowable dose for radiation workers should be a five-year average of 20 mSv per year (100 mSv over 5 years), with a maximum of 50 mSv in any one year. The overall average dose for workers occupationally exposed to artificial sources is less than 1 mSv per year. This average is slightly higher for nuclear workers than for health care workers. For the general public, the five-year average is 1 mSv per year. Exposure in both categories is above background levels and does not include medical exposure. In situations where the public is expected to be affected, the additional dose compared to the background dose received by the public is controlled (Zakariya and Kahn, 2014).

During medical diagnosis and treatment, both medical personnel and patients are exposed to radiation. Radiation protection guidelines also state that radiographers should handle radiation equipment. However, in some applications, doctors or nurses may be present with the radiographers. Medical personnel working with radiation sometimes have the same risk of radiation exposure as the patient in treatment or diagnostic rooms. Since nurses are involved in some diagnoses and treatments, radiation exposure is inevitable. Employees responsible for maintaining medical devices may be exposed to medical radiation (Feng et al., 2022). In addition to medical use, workers in many sectors around the world interact with radiation (Figure 5).

4. Radiation Awareness

Awareness is a distinctively open and receptive form of consciousness in which stimuli are not evaluated, classified or analysed. In awareness, the approach to immediate experience has qualities of openness, acceptance and impartiality. There is an orientation to experience and it is completely non-judgmental (Dimidjian and Linehan, 2003). Awareness is the ability of a person to discriminate what and how he or she is experiencing when interacting with another person through the senses. Through awareness, people's awareness of themselves, their environment and the universe increases, and their field of consciousness expands. For any stimulus to be perceived, it must first exceed a certain threshold and be strong enough to reach the sensory organs. However, one may not be aware of everything that reaches the sense organs. In mindfulness, attention is focused on oneself without judgement. Thoughts, emotions and physical sensations are observed without judgement and as normal and temporary parts of immediate experience (Thompson and Waltz, 2007). Mindfulness is the ability to focus attention and focus on the present moment without judgement (Zinn, 2003). Awareness is the fundamental element in the formation of the life cycle or the relationship between the individual and the environment.

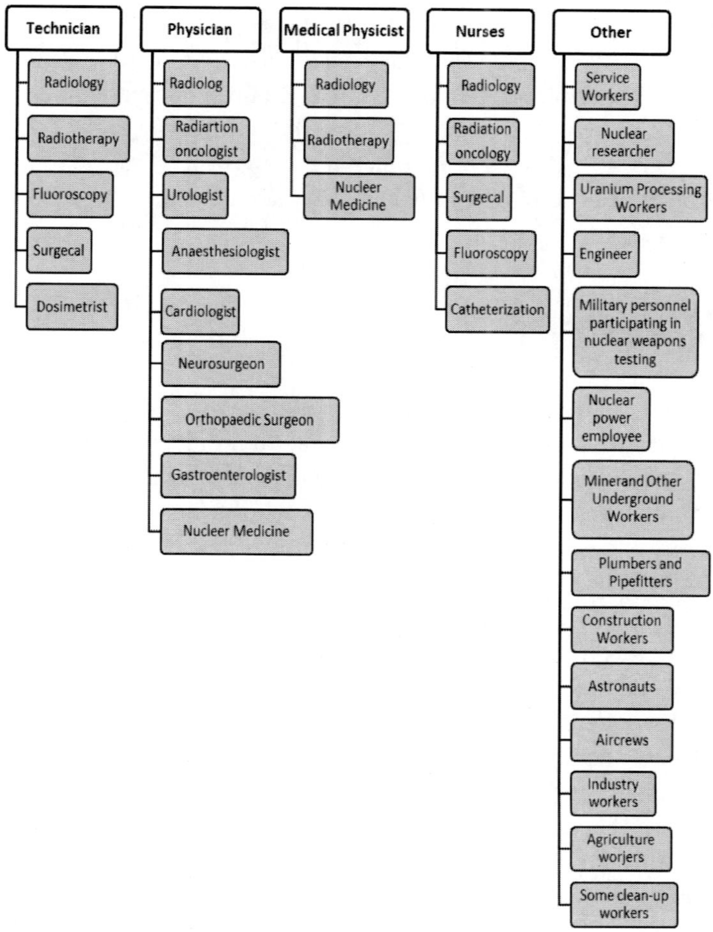

Figure 5. Workers exposed to radiation in the workplace.

Humans have had to live with radiation since the creation of the Earth. The long-lived radioactive elements that have been present in nature since the beginning of time create a natural level of radiation on the Earth's surface that is considered normal and unavoidable. (O'Sullivan et al., 2010). In the last century, this natural level has increased significantly due to nuclear bomb testing and the use of some technological products. Many factors determine the level of natural radiation exposure: where people live, soil structure, materials used in buildings, seasons, distance from poles and weather conditions are some of these reasons. In addition, people receive doses from artificial sources, which are inevitably used in many areas such as energy production, medicine, industry, research, agriculture and animal husbandry (Tack and Gevenois, 2004). Although people's living standards vary according to the physical characteristics of the environments in which they live and geographical conditions, they can be exposed to an annual dose of about 2.1 mSv per person worldwide (Hendry et al., 2009). Social awareness is the driving force behind governance, regulation and law enforcement in general. It can also be a defence against the adverse effects ofradiation.

4.1. Public Radiation Awareness

People need to interact with other people to develop their social lives. When they interact with their cultural and social environment, they develop both socially and culturally and can increase their awareness through education and communication. Awareness increases the power to interact with events in the environment. Society is exposed to various sources of ionising radiation in everyday life, both natural and man-made. Ionising radiation is a widespread, somewhat complex and often misunderstood issue. Exposure to ionising radiation increases the likelihood of both acute and chronic diseases, especially as the radiation dose increases (Christodouleas et al., 2011; Linet et al., 2012). The average radiation dose to the public is 2.5 mSv per year, of which 15% is medical exposure (Wootton, 1991).

As radiation is an abstract concept, the public may find it difficult to comment on the issue (Alsop, Hanson and Watts, 1999). Studies show that there are differences in risk perception and knowledge of the actual sources of ionising radiation between the public and radiation experts (Kanda et al., 2012; Slovic, 1996). Public awareness of radiation is influenced by several factors. The public has fears and concerns about radiation for a number of reasons. They arise from the way the mass media present health risks, which can misinform the public, for example by exaggerating some sources and minimising the risks. Media studies of the risks and dangers of radiation exposure have consistently found that most people overestimate the risk from industrial sources and underestimate the risk from medical uses of radiation (Dowd and Steven, 1998). However, radiation exposure from medical applications is always high and carries a risk of cancer. For example, people who underwent multiple CT scans over approximately 9.5 years had a 24% increase in cancer incidence (Hill and Einstein, 2015).

The issue of radiation risk assessment plays an important role, especially given the educational differences in society (Perco, 2014). In reality, the perception of the risk of nuclear power is highly emotional and difficult to overcome. The public generally perceives nuclear energy as extremely risky, and the risk of exposure to nuclear energy is much higher than it actually is (Kivimäki, Kalimo, 1993). Another factor contributing to the widespread fear of radiation is the lack of reliable and accessible information and the resulting misunderstandings. After the atomic bombs dropped on Hiroshima and Nagasaki and the explosion at the Chornobyl nuclear power plant, the issue of radiation as a major health hazard was widely discussed. As a result, the films that have been made, the information in the literature and the news in the media also support this negative image (Tada, 1999). Radiation has both positive and negative effects on human life. Proper public awareness of radiation will strengthen the approach in both positive and negative aspects.

4.2. Radiation Worker Radiation Awareness

Every worker wants a healthy and safe workplace. Radiation workers may not be sufficiently aware of the control of working time in the workplace and of personal protective clothing or special protective equipment. They may not be aware of the problems that can arise from malfunctions in the use of radiation protection equipment. In some cases, the number of radiation protection devices may be increased out of unnecessary fear and they may feel uncomfortable in their work areas.

Radiation workers are people who are exposed to radiation as a result of their work. There are about 800,000 workers in the nuclear industry and an estimated 2.3 million health workers worldwide. Half of the radiation to which these people are exposed is man-made artificial radiation and ionising radiation (UNESCEAR, 2000). The exposure of workers in different occupations to ionising radiation has become an important topic of study since the second half of the 20th century. Studies have looked at cancer incidence and mortality in people exposed to ionising radiation in their working environment, such as nuclear industry workers, radiologists, radiotherapists and military personnel involved in nuclear weapons testing. These studies have focused on the importance of radiation protection, particularly for long-term low-dose exposures (Vaiserman et al., 2018).

With the increase in radiation sources, many studies have been carried out to determine the radiation awareness of radiation workers, whose exposure is much higher than that of the general public. For example, the lack of radiation awareness among healthcare workers has negative consequences for themselves and their patients. For example, the demand for radiological applications sometimes increases uncontrollably. It has become a subject of research both for the physician who requests the examination, the technologist who performs it, and the patient who is exposed to radiation (Hricak et al., 2011; Laure, 2009). Recently, there has been increasing awareness of the lack of knowledge among referring physicians about the radiation doses exposed during diagnostic radiology procedures (Azmoonfar et al., 2016).

The 2010 study by Luk, Leung, and Cheng aimed to determine the awareness of radiation among young clinicians. For this purpose, questions were prepared and asked to measure the level of knowledge about radiation, and it was found that young doctors had insufficient knowledge about this topic. Kam's study, which involved 130 doctors in Hong Kong in 2005, aimed to determine doctors' awareness of radiation doses. The results showed that their knowledge and awareness of radiation was inadequate. Healthcare workers who work with radiation may be at high risk of radiation exposure in treatment or diagnostic rooms. Quinn et al. (1997) reported that in the UK, people without radiation training do not have a clear awareness of radiation protection. Faggioni et al. (2017) investigated the medical radiation awareness of medical students, radiography students and radiology assistants in Italy. They found that the radiation protection awareness performance of medical students was significantly lower than that of radiography students and radiology assistants.

A study by Samuel Amin Sampong found that nurses were not aware of the harms of radiological examinations and how to protect themselves (Anim Sam et al., 2015). Another study conducted in the Najran region found that health workers were not sufficiently aware of the risks and protection from ionising radiation, but after one week of training, awareness increased sufficiently (Asiri, 2023).

It is a fact that people who have received radiation training or work in radiation-related jobs have higher knowledge indexes (Feng et al., 2022). In fact, radiation workers should be trained on radiation and its effects on health, radiation protection and safe use before starting work, and the importance of protective equipment and dosimetry should be emphasised. Dose monitoring should be carried out regularly. Following dosimetric measurements, personnel with dose exceedances should be allowed to take time off. The use of dosimeters by health care workers should be encouraged through in-service training programmes, and confidence in dosimeter results should be increased.

Conclusion

Although radiation protection is defined by international organisations, the public may not have sufficient knowledge about radiation and radiation protection. For example, they may not be aware that there is natural radiation in food or water. Given the importance of accurate radiation information and the development of technology, it is a matter of social responsibility to be aware of this issue, which affects many people. As society and radiation workers become more aware, their perspectives on events in the environment may change. Radiation has become a topic of interest, especially as children and young people will be using this technology extensively, especially due to the increase in man-made radiation sources as technology develops.

Because radiation cannot be seen, smelled, tasted or felt, the level of radiation awareness of the general public and radiation workers can be very different. The issue of radiation in our daily lives is too important to ignore. Although the application of medical, scientific and industrial practices for the benefit of mankind is important, it is necessary to be informed about the risks of radiation exposure. Many studies show the importance of radiation awareness. A better understanding of radiation can improve the public's ability to recognise the reality of radiation-related issues, to protect themselves from the potential of radiation risks, and to contribute positively to the sustainable development of society. Continuing medical education in radiation protection practices is necessary to raise awareness among those occupationally exposed to ionising radiation. There is a need to improve public engagement and education on radiation and risk. Healthcare providers and the Internet are identified as the preferred educational resources for the public. In the light of scientific knowledge, the harmful and beneficial aspects of radiation that are commonly considered can also be brought to the fore.

The public and health professionals need to be fully informed about the dangers and precautions of radiation exposure so that they can protect their health by increasing radiation awareness. Providing up-to-date information on the importance of radiation technology and its applications can increase students' interest in the subject. Providing access to accurate information and reinforcing the subject through the use of computers and the Internet can increase awareness.

References

Alsop, S., Hanson, J., and Watts M. (1999). Pupils' perceptions of radiation and radioactivity: The wary meet the unsavoury, *School Science Review,* 79,75–79.

Anim-Sampong, S., Opoku, S. Y., Addo, P., & Botwe, B. O. (2015). Nurses knowledge of ionizing radiation and radiation protection during mobile radiodiagnostic examinations. *Educ Res,* 6(3), 39-49.

Asiri, A. A. M. (2023). Increasing awareness of radiation hazard and radiation protection among medical staff. *Acta Scientiarum. Health Sciences,* 45.

Attix, F. H. (2008). *Introduction to radiological physics and radiation dosimetry*. John Wiley & Sons. Weinheim, Germany.

Azmoonfar, R., Faghirnavaz, H., Younesi, H., Morovati, E., Ghorbani, Z. H., & Tohidnia, M. (2016). Physicians' knowledge about radiation dose in radiological investigation in Iran. *Journal of biomedical physics & engineering,* 6(4), 285.

Brenner, D. J., Doll, R., Goodhead, D. T., Hall, E. J., Land, C. E., Little, J. B., ... & Zaider, M. (2003). Cancer risks attributable to low doses of ionizing radiation : assessing what we really know. *Proceedings of the National Academy of Sciences,* 100(24), 13761-13766.

Christodouleas, J. P., Forrest, R. D., Ainsley, C. G., Tochner, Z., Hahn, S. M., & Glatstein, E. (2011). Short-term and long-term health risks of nuclear-power-plant accidents. *New England journal of medicine,* 364(24), 2334-2341.

Cupitt, J. M., Vinayagam, S., & McConachie, I. (2001). Radiation exposure of nurses on an intensive care unit. *Anaesthesia,* 56(2), 183-183.

Dehghani, A., Ranjbarian, M., Mohammadi, A., Soleiman-Zade, M., & Dadashpour-Ahangar, A. (2014). Radiation safety awareness amongst staff and patients in the hospitals. *International journal of occupational hygiene,* 6(3), 114-119.

Dimidjian, S., and Linehan, M. M. (2003). Defining an agenda for future research on the clinical application of mindfulness practice. *Clinical Psychology: Science and Practice,* 10 (2): 166-171.

Dowd, S. B., & Steves, A. M. (1998). *Patient education: Communicating radiation risk.* ASRT.

Faggioni, L., Paolicchi, F., Bastiani, L., Guido, D., & Caramella, D. (2017). Awareness of radiation protection and dose levels of imaging procedures among medical students, radiography students, and radiology residents at an academic hospital: results of a comprehensive survey. *European journal of radiology,* 86, 135-142.

Feng, C. J., Lai, Y. C., Lee, S. H., Lien, K. Y., Tseng, C. Y., Wu, N. S., ... & Hsu, S. M. (2022). The Analysis with Quantitative Indexes for Public's Awareness of Radiation Knowledge in Taiwan. *International journal of environmental research and public health,* 19(20), 13422.

Feng, C. J., Lai, Y. C., Lee, S. H., Lien, K. Y., Tseng, C. Y., Wu, N. S., ... & Hsu, S. M. (2022). The Analysis with Quantitative Indexes for Public's Awareness of Radiation Knowledge in Taiwan. *International journal of environmental research and public health,* 19(20), 13422.

Frischknecht, R., Braunschweig, A., Hofstetter, P., & Suter, P. (2000). Human health damages due to ionising radiation in life cycle impact assessment. *Environmental impact assessment Review,* 20(2), 159-189.

Glasser, O. (1995). WC Roentgen and the discovery of the Roentgen rays. *AJR. American journal of roentgenology,* 165(5), 1033-1040.

Havas, M. (2017). When theory and observation collide : Can non-ionizing radiation cause cancer?. *Environmental Pollution,* 221, 501-505.

Hendry, J. H., Simon, S. L., Wojcik, A., Sohrabi, M., Burkart, W., Cardis, E., & Hayata, I. (2009). Human exposure to high natural background radiation : what can it teach us about radiation risks?. *Journal of Radiological Protection,* 29(2A), A29.

Hill, K. D., & Einstein, A. J. (2016). New approaches to reduce radiation exposure. *Trends in cardiovascular medicine,* 26(1), 55-65.

Hricak, H., Brenner, D. J., Adelstein, S. J., Frush, D. P., Hall, E. J., Howell, R. W., & Wagner, L. K. (2011). Managing radiation use in medical imaging: a multifaceted challenge. *Radiology,* 258(3), 889-905.

Kam, C. K. (2005). A survey on doctors' awareness and attitude of radiation dose of imaging examination in Hong Kong. *Postgraduate Thesis.* The University of Hong Kong, Medical Sciences, Hong Kong.

Kanda, R., Tsuji, S., & Yonehara, H. (2012). Perceived risk of nuclear power and other risks during the last 25 years in Japan. *Health physics,* 102(4), 384-390.

Keith, S., Doyle, J. R., Harper, C., Mumtaz, M., Tarrago, O., Wohlers, D. W., & Barber, L. E. (2012). Toxicological profile for radon. Atlanta (GA): agency for toxic substances and disease registry (US). Appendix D, overview of basic radiation physics. *Chemistry, and Biology.* Available from: https://www.ncbi.nlm.nih.gov/books/NBK158792/.Accessed May, 21, 2022.

Kivimäki, M., & Kalimo, R. (1993). Risk perception among nuclear power plant personnel: a survey. *Risk Analysis,* 13(4), 421-424.

Lauer, M. S. (2009). Elements of danger—the case of medical imaging. *Minnesota Medicine,* 92(12), 40.

Linet, M. S., Slovis, T. L., Miller, D. L., Kleinerman, R., Lee, C., Rajaraman, P., & Berrington de Gonzalez, A. (2012). Cancer risks associated with external radiation from diagnostic imaging procedures. *C A: a cancer journal for clinicians,* 62(2), 75-100.

López P. O., Dauer L. T., Loose R., Martin C. J., Miller D. L., Vañó E., Doruff M., Padovani R., Massera G., Yoder C., Authors on Behalf of ICRP. ICRP Publication 139 : Occupational Radiological Protection in Interventional Procedures. *Ann ICRP.* 2018 Mar; 47(2):1-118. [PubMed].

Luk, S. Y., Leung, J. L. Y., & Cheng, C. S. (2010). Knowledge of radiation dose and awareness of risks : a cross-sectional survey of junior clinicians. *Hong Kong Journal of Radiology,* 13(4), 189.

Mehta, S. R., Suhag, V., Semwal, M., & Sharma, N. (2010). Radiotherapy: basic concepts and recent advances. *Medical Journal Armed Forces India,* 66(2), 158-162.

Mehta, K. (2005). Radiation: basic principles. *Journal of vascular surgery,* 42(6), 1237-1238.

Mettler Jr, F. A., Mahesh, M., Bhargavan-Chatfield, M., Chambers, C. E., Elee, J. G., Frush, D. P., ... & Vetter, R. J. (2020). Patient exposure from radiologic and nuclear medicine procedures in the United States : procedure volume and effective dose for the period 2006–2016. *Radiology,* 295(2), 418-427.

Mettler, F. A., & Upton, A. C. (2008). *Medical effects of ionizing radiation.* Saunders/Elsevier.

O'Sullivan, J., O'Connor, J. and O'Regan K. (2010). An assessment of medical students awareness of radiation exposures associated with diagnostic imagingin vestigations. *Insights Imaging,* 1 : 86-92.

Perko, T. (2014). Radiation risk perception : a discrepancy between the experts and the general population. *Journal of environmental radioactivity,* 133, 86-91.

Podgorsak, E. B. (2005). (Editor), *Radiation Oncology Physics: A Handbook for Teachers and Students, International Atomic Energy Agency (IAEA),* Vienna, Austria, 2005.

Quinn, A. D., Taylor, C. G., Sabharwal, T., & Sikdar, T. (1997). Radiation protection awareness in non-radiologists. *The British journal of radiology,* 70(829), 102-106.

Ramanathan, S., & Ryan, J. (2015). Radiation awareness among radiology residents, technologists, fellows and staff : where do we stand? *Insights into imaging,* 6, 133-139.

Rosen, E. M., Day, R., & Singh, V. K. (2015). New approaches to radiation protection. *Frontiers in oncology,* 4, 381.

Slovic, P. (1996). Perception of risk from radiation. Radiation protection dosimetry, 68(3-4), 165-180.

Tack, D., & Gevenois, P. A. (2004). Radiation dose in computed tomography of the chest. *Jbr-Btr,* 87(6), 281-288.

Tada, J. (1999). Understanding radiation and risk: The importance of primar and secondary education. Japan Nippon Genshiryoku Kenkyujo JAERI, Conf 399 - 403.

Thompson, B. L.and Waltz, J. (2007). Everyday mindfulness and mindfulness meditation : Overlapping constructs or not? *Personality and Individual Differences,* 43(7), 1875-1885.

UNSCEAR, (2000). Report on Sources and Effects of Ionizing Radiation to the General Assembly (2 Volumes), United Nations, Vienna 2000.

Unscear, S. (2000). Sourses and effects of Ionizing Radiation. United Nations, New York, 453-487.

Vaiserman, A., Koliada, A., Zabuga, O., & Socol, Y. (2018). Health impacts of low-dose ionizing radiation : current scientific debates and regulatory issues. *Dose-Response,* 16(3), 1559325818796331.1.

WHO. 2023. Available from https://www.who.int/news-room/fact-sheets/detail/ionizing-radiation-and-health-effects.

Wootton, R. (1991). The POPUMET regulations : careless radiology costs lives. *British journal of hospital medicine,* 45(3), 133.

Zakariya, N. I., & Kahn, M. T. E. (2014). Benefits and biological effects of ionizing radiation. *Sch. Acad. J. Biosci,* 2(9), 583-591.

Zinn, K. J. (2003). Mindfulness based interventions in context : Past, present, and future. Clinical Psychology: *Science and Practice,* 10(2): 144-156.

Chapter 11

Exposure and Dose Rate in Radiation Accidents

Nilsu Çini*

Department of Radiation Oncology, Kartal Dr. Lütfi Kırdar City Hospital, Istanbul, Türkiye

Abstract

A radiation accident could be defined as an incident involving an unplanned or unforeseen radiation emission, misuse of radiation devices, or misapplication or mislay of radioactive substances. A radiation accident could be related to a broad spectrum of applications or utilisation of radiation sources, including industry, medicine, and nuclear facilities. This widespread use will have more communal and environmental outcomes after a critical accident than expected. Medical and environmental intervention schemes and schedules need to be planned based on calculated severity estimations of the event. The accident severity will be determined based on the radiation dose rate and dose exposure measurements and calculations done on-site and on time or retrospectively. The intensity and type of radiation are key factors in determining the health effects of radiation exposure. Accident dosimetry is a vital process that must be done quickly and with the most appropriate techniques for the type of accident. Rapid recording, collection and evaluation of accident-related data will accelerate the intervention of response organisations in determining what to do in the short term and long term after the event. Additionally, it will provide data for post-accident dose reconstruction. The dose rate report indicating the severity and magnitude of exposure determined by the

* Corresponding Author's Email: nilsucini@gmail.com.

In: Triage in Radiation Accidents
Editors: Nural Öztürk and Sibel Karaca
ISBN: 979-8-89530-415-0

fast and highly accurate dosimetric techniques is a must for triage of exposed people and decontaminating the environment following the accident. It will be the most important part of the entire process management. To reveal the whole process, in this section, we will try to mention physically and biologically based retrospective dosimetric techniques for accident dosimetry and post-accident dose reconstruction, as well as exposed dose estimations based on acute radiation syndrome symptoms.

Keywords: acute radiation syndrome, dose rate, exposure, radiation types, retrospective accident dosimetry

1. Introduction

A radiation accident is an important event that cannot be predicted, requires urgent intervention, and has effects on people and the environment at different levels. Radiation is a form of energy that physically interacts with matter as it moves through the medium. The number of effects caused by these physical interactions due to radiation exposure can be estimated by dose measurements. The biological effect of radiation exposure on the organism is dose rate. The risk to a particular organ or tissue depends on the average energy absorbed through the exposed target. Additionally, the biological consequences of exposure are affected by the total dose absorbed, as well as the type of ionising radiation that causes energy deposition, duration of exposure, and dose rate (Rühm et al., 2015).

Measurement of the ionising radiation doses that exposed individuals may receive during and after a nuclear or radiological accident is crucial in managing the aftermath of such an adverse event. To make decisions regarding health and life protective actions and create a triage route, it is necessary to use effective dose measurement techniques while making reliable and precise dose estimates for the employees and society affected by the accident. It is essential to determine the short- and long-term health effects that may occur in the public and occupationally exposed individuals following exposure to ionising radiation due to nuclear and radiological accidents (Barquinero et al., 2021).

During a radiation accident, both occupationally exposed people and members of the public may be in the environment. Information can be obtained from workers' dosimeters or radiation field monitors for rapid exposure estimates. However, in radiation accidents that may cause very high radiation exposure, these devices may be insufficient or overdose. Members of the public do not have dosimeters anyway, so additional information is needed.

In line with this need, retrospective physical and biological dose measurement techniques have been developed for accident dosimetry. This chapter will discuss several dose measurement techniques developed using substances in the environment, materials on the exposed person, or biological materials of the exposed individual, as well as the strengths, weaknesses, and limitations of these techniques. In addition, the importance of the type of exposure, amount of radiation, and time elapsed after exposure will be considered in choosing the suitability of these retrospective dose measurement techniques.

In accident dosimetry, it is momentous to obtain precise dose estimates by choosing or combining one of the biological or physical retrospective dosimetry techniques according to these differences. It should be underlined that exposure and dose rate measurements in radiation accidents are vital procedures that must be done and well documented.

2. Types, Sources and Fields of Use of Radiation

Radiation is an energy form. Radiation energy is emitted by a source and then travels through a medium such as air, the human body, water, or an apparatus while making physical interactions until it is absorbed by matter (Galindo, 2023). Electromagnetic radiation and particle radiation are the forms of radiation (Tuieng et al., 2021). Nuclear particles, like high-speed electrons, protons, and neutrons that cause ionisation effects are particle radiation (Percuoco, 2014). Electromagnetic radiation species include gamma rays and x-rays in addition to radio waves, microwaves, infrared, and ultraviolet. Electromagnetic radiation is categorised as non-ionising and ionising types. Both types can be used clinically or environmentally, and exposure may potentially affect organisms positively or negatively (Tuieng et al., 2021).

Ionising radiation has enough energy to remove an orbital electron from a stable atom or molecule. This behaviour can be very harmful to the staminal processes of organisms, inducing DNA damage that underlies a variety of life-threatening situations, including cancer. Generation of reactive oxygen species and ionisation leads to DNA double-strand breaks (DSB) and/or single-strand breaks (SSB), and interstrand cross-linkages on the cellular level. The cellular repair mechanism may not be able to properly or completely repair DNA damage that causes cell death or transformation, depending on the type, dose rate, and total dose of ionising radiation (Goodarzi et al., 2016).

Techniques developed to measure damage levels in DNA are the basis of retrospective biodosimetry. Biodosimetry is mostly based on various cytogenetic analyses that can detect DNA damage and misrepair radiation exposure induced. Also, the induction of chromosome aberrations in human peripheral blood lymphocytes is one of the most useful dosimetry systems for accidental whole-body irradiation (Zoetelief & Broerse, 1990). We will look after these techniques in Section 4.

Overall, the worldwide average effective dose rate from natural background is about 2.4 mSv/annual (Thorne, 2003). Although all organisms and humans are exposed to background radiation, exposure to man-made sources of radiation differs for the following groups: Members of the Public and Occupationally Exposed Individuals. Therefore, the International Commission on Radiological Protection (ICRP) defined and declared dose limitations in the 2007 Recommendations report (International Commission on Radiological Protection [ICRP], 2007). In the ICRP report dose limits are split into two groups for the public and occupationally exposed workers, see Table 1. According to ICRP, occupational exposure refers to all exposure incurred by workers during their shift. But exposure from exempt activities involving radiation, any medical exposure or exposure to normal local natural background radiation are except for these limitations.

Table 1. Dose limits table for occupationally exposed workers and the public (International Commission on Radiological Protection [ICRP], 2007)

	Stochastic Limits Effective Dose	Deterministic Limits Annual Equivalent Dose in Skin	Deterministic Limits Annual Equivalent dose in the Lens of the eye
Occupational	20 mSv per year, averaged over a defined 5 years	500 mSv	150 mSv
Public	1 mSv in a year	50 mSv	15 mSv

Limitations given in Table 1. as Sievert (Sv) unit. A Sv is the unit of effective dose, and also the unit of equivalent radiation dose. One Sv is the amount of radiation that is necessary to produce the same bioeffect on tissue as 1 Gray (Gy) of high-energy X-rays. Quantities that are measured in the Sv unit indicate the ionising radiation biological effects. The energy absorbed by 1 kg of biological tissue is 1 Sv and has the same effect as 1 Gy absorbed dose of gamma radiation. Sievert can also be expressed as 1 Sv = 1 Joule/kg. Gray is the unit of absorbed dose and specific energy per unit mass. The specific energies representing this type are usually gamma particles or X-rays. Its definition is the energy absorption of 1 joule of ionising type radiation by 1 kg of tissue. From this definition, a Gy is expressed as $m^2 \cdot s^{-2}$ (Cameron, 1991).

As a result of such various uses of radiation, many precautions have been taken and protocols stating radiation dose limitations and radiation protection methods have been published. Despite these precautions, we are always faced with a significant possibility of a radiation accident. After a radiation accident, how accidental dose exposure will be managed?

3. Radiation Accidents

In the general population radiation injury developed due to radiation incidents or accidents is a very rare but remarkable event (Cerezo, 2011). Radiation accidents occur in medical, industrial, and research settings, where radioactive materials or radiation-generating equipment (such as X-ray machines or particle accelerators) are used (George A. Alexander, 2016).

Another definition of an accident by the IAEA is, "an event that has led to significant consequences to people, the environment or the facility"(Cerezo, 2011). If we look back at radiation accidents, there have been 600 accidents reported since 1945, resulting in more than 200 deaths. Of these reported accidents, 4% military, 11% nuclear, 11% medical, 19% research, 41% industrial, and 14% were unidentified. External exposure was 84%, of which 12% was whole body and the rest was localised. Internal exposure was 10% and the remaining 6% was mixed. Chernobyl and Fukushima are not included in this analysis (IAEA, 2024). Of the total energy released in a nuclear explosion, 35% is heat, 50% is shock or bomb explosion, and the remaining 15% is radiation emission (López & Martín, 2011).

Externally irradiation exposure from a source distant to or near the body is an external exposure accident. Radiation exposure stops when a person is removed from that radiation source, or the radiation generated machine is closed. Individuals exposed externally do not become radioactive and do not cause danger to other people around. A contamination accident is a type of accident in which radioactive material contaminates the individual and the radioactive exposure will continue until the contamination material is removed or completely decays. Radioactive particles, liquids, or gases are the forms of possible contamination. Contamination can spread both to other parts of the individual's own body and to other individuals in the environment. The third type of accident is a combination of these two types (George A. Alexander, 2016).

Radiation workers and the public may be irradiated during accidents. There are two groups of people that we can talk about after a radiation accident except the emergency workers. The first group is radiation workers who are occupationally exposed and have dosimeters while working, and possibly during the radiation accident. However, since such accidents involve uncontrolled and may be high exposure, the radiation found in the environment can affect patients, and the public as well as the workers, who are not occupationally exposed. This second group of people does not have a dosimeter. Determining post-accident exposure of the second group will be an important issue and may need some special methods. In a radiation accident emergency, authorities in accident response organisations should take all measurements to record and evaluate the radiation exposures of workers. Radiation exposure evaluations of the emergency workers should be monitored individually, like direct reading, alarm dosimeter, etc.

The severity of potential damage that may occur after excessive radiation exposure depends on various factors that affect the calculation of the exposure dose. These factors include the type of radiation, the sensitivity of the affected tissues and organs to radiation, the route, distance and duration of exposure, and the individual characteristics of the exposed person (such as age, gender and general health status). Dosimetric measurements are very important for establishing dose-side effect correlation and realistic damage prediction (WHO, 2023).

Dosimetry in accidents is a need, especially for medical purposes. First, initial-phase dose assessment is vital to identify who needs medical intervention due to deterministic tissue effects. In Table 2 you can see the possible biological effects of short-term radiation exposure correlated magnitude of exposure in absorbed dose unit Gy(López & Martín, 2011).

The evaluated and recorded doses could be a part of epidemiologic and long-term risk determination. Dosimetric data obtained long after the event will provide relevant stochastic injury information for epidemiological studies. Dosimetric data obtained shortly after the event will maintain relevant deterministic tissue damage information. The initial dose exposure assessments are necessary for the rapid guidance to health experts in electing those waiting for immediate triage and medical treatment, and prevent redundant overloading of medical facilities with individuals exposed lightly. Also, it is important to determine the individual or environmental exposure type, whether it is partial or whole-body irradiation (IAEA, 2024).

Table 2. Biological effects of short-term radiation exposure on humans

Acute Effect	Dose
No detectable effects.	< 0,2 Gy
A temporary decrease in white blood cell count, defined as measurable temporary blood changes.	0,2-1 Gy
Nausea, vomiting and long-term decrease in white blood cells, defined as acute radiation sickness.	1-2 Gy
Vomiting, weakness, loss of appetite, diarrhoea, death in some cases.	2-3 Gy
Vomiting, diarrhoea, haemorrhaging, and death without medical treatment in 50% of cases at exposed 3.5 Gy or above.	3-6 Gy
Eventual death in almost all cases.	> 6 Gy

Ionising radiation lethality is determined by a number of factors. First, dose rate: doses taken in a relatively short period of time will cause more harm. Second, distance from the source, the dose rate decreases proportionally to the square of the increasing distance (inverse square law-for point radiation sources). Third, shielding may reduce the exposed dose depending on the material used on the exposed surface and the type of radiation (Reeves, 1999). Fourth, for people who have been exposed to moderately high radiation doses, access to medical therapy is essential (López & Martín, 2011). The correlation between dose rate and exposure rate is, that the total ionising radiation per hour in a person's circle is exposure rate (mR/h, milliRoentgen per hour), and the biological effect on the body originating from exposure to that radiation is dose rate (nSv/h, nanoSieverts per hour).

The International Nuclear and Radiological Event Scale (INES) classifies radiation events on seven levels depending on their magnitude: levels 4–7 are "accidents," and levels 1–3 are named "incidents." A logarithmic approach has been applied in this scaling; with each increase in the scale level, the severity of the event is approximately ten times greater. Events are scored in terms of their impact on people and the environment. Events that do not pose a safety hazard are scored as Level 0 (IAEA, 2024). The Chernobyl reactor explosion, exposure levels greater than 2 Gy occurred due to the external release of massive nuclear products, is a level 7 accident on this scoring. After this accident, it was reported that out of approximately 150 workers diagnosed with acute radiation syndrome (ARS), 30 died in the early period and 14 died within the next 10 years (LiVolsi et al., 2011).

4. Accident Dosimetry

Following accidental exposure to high doses, information on the absorbed dose by various tissues and organs is required. The manifestation of acute effects depends primarily on the absorbed dose by the tissue rather than on shielding or other radiation-related quantities. On the other hand, radiation weighting factors do not apply here as they are the subject of years of occupational exposure. Biological acute effects are also affected by tissue/ organ specificity and Relative Biological Effectiveness (RBE) value specific to those organs. Organs represent different responses to radiation exposure (Barazzuol et al., 2020). The side effects of radiotherapy in normal tissue can be divided into early (or acute) and late responses, depending mostly on tissue turnover time and their modulation by processes that mimic a wound-healing response (Bentzen, 2006).

Let's not forget that the response to exposure can also be individual and some people may show these effects faster, others may not. This is the issue of interpatient intrinsic radiosensitivity (West, 1995). The fact that these effects do not occur in some of the people who were in the same environment at the time of the accident, does not mean that those people were not exposed; a certain prophylactic treatment or triage should still be applied. We will talk about the relationship between ARS symptoms, exposure dose and triage in section 5.

In post-accident dose estimations for individuals, exposure amounts should be presented in the "absorbed dose" unit Gray in the initial dose assessment report. The Gy unit should be used in all cases to avoid confusion and distrust that may arise from the use of multiple and different units in reports created from different centres (IAEA, 2024).

Dosimeter selection may vary depending on the kind of radiation to be evaluated. For the evaluation of D_T in tissues and organs, dosimeters for photons and neutrons provide information about the personal single dose equivalent Hp (10), Hp (10) is a dose equivalent at 10 mm depth in tissue, also known as whole body dose (Shubayr & Alashban, 2021). Dosimeters for eye lenses provide info on Hp (3) for photon and beta radiations (Tanaka et al., 2019). Extremity dosimeters provide information about skin dose. Dose values ranging from 250 mGy to 10 Gy generally matter and 100 mGy is the lower interest limit. As a special case, for highly non-uniform irradiations the dose rate can be up to 10^5 Gy/s.

Regular personal dosimeters should be able to measure up to 10 Gy to eliminate the need for an additional dedicated accident dosimeter. Some types of dosimeters, such as film, cannot provide dose data over a wide energy range. Therefore, regular use of dose rate meters or warning alarm dosimeters may be effective in avoiding significant exposures. Especially, used for high dose-rate fields, dosimeters should be very reliable. In accidents involving fissile products, additional neutron dosimeter equipment is needed for accident dosimetry measurements.

The dosimetry system has to have some capabilities for immediate triage and give some outputs. Many persons may have been exposed but some people require immediate medical attention. Therefore, the dosimetry system must provide an exposure indicator to estimate maximum doses, assess doses for everyone exposed, and identify exposed persons for a rapid initial screening. Another dosimetry may be performed for contamination which presumably is a concern.

During the dose reconstruction, multiple inputs and data obtained from dosimetric measurements were used to sense the whole incident. Evaluation of post-accident exposures can first start using the dose data obtained from working area monitors and the data about the location of the radiation source. Then, the location of exposed persons relative to the radiation source, presence of any shielding, duration of exposure, and physical properties of the radiation field in the environment in which they are located are added. Additionally, data on irradiation uniformity and spectral dose distribution over the exposed individual body should be recorded. If data is obtained about the accident period not certain, retrospective dosimetry, accident simulation and mathematical dose reconstruction techniques are used.

4.1. Dose Rate Measurements in Radiation Accidents

The biological effect of radiation exposure on the organism is dose rate (nSv/h). The role of dose rate is an issue that still contains significant uncertainties, despite research on the bioeffects of ionising radiation (Lowe et al., 2022).

The risk to a particular organ or tissue depends on the average energy absorbed through the exposed target. Additionally, the biological consequences of exposure are affected by the total dose absorbed, as well as the type of ionising radiation that causes energy deposition, duration of exposure, and dose rate (Rühm et al., 2015). Extrapolation of dose data output at high doses and high dose rates down to low doses and low dose rates is needed, to determine radiation hazards in accidental scenarios relevant to radiation protection (Rühm et al., 2018).

Sufficient quantities of suitable dose rate meters capable of providing a direct reading of the environmental dose equivalent rate H (10)/h in the range of μSv/h to 1 Sv/h should be available for the dose rate measurements in high exposure operations or accidents. Equipment calibrated to H'(0.07)/h should be available in accidents where high levels of surface contamination by beta/gamma emitters are anticipated. Appropriate energy range, durability, battery life, and usable temperature range are important criteria for these dosimeters.

5. Retrospective Dosimetry: Retrospective Techniques for Accident Dosimetry

Dosimeters belonging to radiation workers, radiation field monitors, and any other dosimetric equipment that exists during the accident should be processed immediately and reported. What if no dosimeter is present during an accident in a public area, for example, a stolen or a missing source exposed to a few people? Several techniques exist when no dosimetric information is possible, dosimetry was damaged or not worn by the exposed individuals. These techniques are called "retrospective dosimetry": In post-accident situations where conventional dosimeters are unavailable or inadequate and, dose assessment of exposed persons is required. Retrospective techniques for accident dosimetry could be divided into two physical dosimetry and biological dosimetry.

Physical dosimetry can be applied to either biological materials, such as teeth and nails, or non-biological materials, such as electronic devices and clothing, of the exposed person. Physical dosimetry procedures measure the dose on the sample appropriate for the physical dosimeter, not for the individual. Therefore, in the dosimetric output report, the sample obtained dose should be converted to the organ dose or 'whole body' dose. Calculations of individual dose exposure are based on information about the distribution, exposure rate, and proximity of the contaminants and also the presence of protection. However, this individual dosimetry is unlikely for a large-scale radiation accident.

Biological dosimetry utilises changes in an organism at the molecular level or cellular level and is most often performed on blood (IAEA, 2024). However, the process of sampling fresh blood and cell cultures from victims using standardised techniques in biological dosimetry is a significant obstacle to rapid and immediate dose reconstruction (Maltar-Strmečki et al., 2022). Biological dosimetry identifies the modified accident dose by RBE. It is the ratio of the reference radiation dose to a specific radiation dose that produces the same bio-effect. In cases where it is possible for the RBE value to be > 1, the linear energy transfer value (LET) of the radiation source is high, whereas for low LET sources RBE $\approx$ 1. Physical dosimetry does not involve the use of the RBE concept as it takes into account the response of the sampled dosimetric material to photon energy.

5.1. Physical Dosimetry

Retrospective Physical dosimetry techniques exhibit affirmative characteristics for a set of materials such as glass, alanine, sugar, plastic, silicates, and tobacco (Maltar-Strmečki et al., 2022). Electron Paramagnetic Resonance (EPR), Thermoluminescence (TL) and Optically Stimulated Luminescence (OSL) are the typical physical dosimetry techniques.

5.1.1. Electron Paramagnetic Resonance (EPR)

The spectroscopic procedure is utilised to identify unpaired electrons in specimens, such as free radicals caused by ionising radiation. Ideally, a sample to be used in EPR dosimetry will be nice to have a limited number of features. They are ubiquitous, non-invasive to collect, easy to prepare, radiation-specific, time-stable, and recognise signals that are not caused by radiation.

Electron paramagnetic resonance on textile fibres/cotton/cellulose allows to define the pattern of the absorbed dose distribution throughout the body and provides easy sample collection. However, there are some confounding factors such as daily exposure to direct sunlight and temporal instability of pre-irradiation signals that can be problematic for EPR. Plastics in many objects in everyday use like watches, mobile phones, credit cards, buttons, and glasses are so easy to collect. Due to several significant disadvantages, such as the huge variety of compositions of the plastic and each type having different properties, plastics are rarely used for accident dosimetry.

Especially in epidemiological studies, the retrospective dosimetry technique EPR on bone and enamel of tooth is considered as the gold standard. In a study conducted with animal tooth enamel in Fukushima, Japan, for the calibration curve the critical dose level was determined as 1.0 Gy where the limit dose for detection was determined as 1.8 Gy. It has been reported that EPR, as a retrospective dosimetric method, will be useful in dose estimation for lifetime dose exposure higher than 1.0 Gy (Harshman & Johnson, 2019). These biological structures provide signals stable enough to retain radiation exposure information for decades. A major limitation is the necessity of tooth extraction for dosimetric measurement, so in vivo techniques are being developed. For very invasive accidents where bone sampling is possible, it is possible to detect long-lived radicals in irradiated bones with EPR spectroscopy.

Hair, human fingernails, and toenails provide signals suitable for EPR dosimetry, and are easy and painless to collect and may be useful for localised exposure. On the other hand, they are imperfect and limited-use tools, nails are not radiation-specific and stable and have high uncertainty. Hair has an elevated background because of its sensitivity to UV irradiation and temperature, the intensity of the signal it contains decreases with time, and characteristics of the hair colour can be an issue. For mineral glass and sugar to be used in EPR is also an option. A tissue-equivalent material sucrose gives a dose-dependent EPR signal, and it is stable. A negligible weak background signal with a dose response up to 10 kGy is linear. Like mobile device touchscreen, mineral glass properties vary between glass types. The highly variable intensity of the background signals and light exposure history gives confounding signals. And a significant degradation of the signal over time following irradiation.

5.1.2. Thermoluminescence (TL) and Optically Stimulated Luminescence (OSL)

In conventional radiation dosimetric measurements, thermoluminescence (TL) and optically stimulated luminescence (OSL) are used for a long period. Emergency response dosimetry uses incidental materials that are on or part of the individual. The positive aspects of this type of dosimeter are high efficiency, high sensitivity, and ease of use. However, accepted uses and standardised protocols have not yet emerged. These types of detectors stock the ionising radiation energy they are exposed to. To evaluate the amount of stocked energy, it is necessary to heat the TL detectors to a temperature of 300°C and release the energy in the form of light. Optically stimulated luminescence dosimeters must be exposed to visible light, encouraging the recombination of trapped electrons and ensuring the emission of optical photons. The resulting optical or luminescent photons are counted, and a curve is given for absorbed dose estimates (Zidouh et al., 2023). Chip cards are used in many electronically enhanced cards and personal data cards. After the chip is removed from the card, it can be used for OSL dosimetry. One study examined three different types of cell phone circuit boards, resistors, tantalum capacitors and ICs, that could be collected after an accident and that reflect the most underlined features of OSL, signal annealing and sensitivity. The OSL dose response of these three components is linear for the dose range investigated (0.7 to 27 Gy for resistors and 0.7 to 160 Gy for capacitors and ICs), suggesting that they can be used as individual OSL exposure dosimeters in the event of a radiation accident. Despite the encouraging results, these ingredients also exhibit strong fading (Bassinet et al., 2010). Based on good evidence, other components integrated into mobile phones should continue to be examined within the framework of this problem. Also, touchscreen glasses of mobile phones are possible to use on TL dosimetry for emergencies due to the radiation-sensitive properties of the glass material. Other personal items like clothing fabrics, plastic cards, jewels, watches, keys, coins, and banknotes, have a high prevalence easy to get the sample. Knowledge regarding the dosimetric properties of these materials continues to develop.

5.2. Biodosimetry

Biodosimetry is mostly based on various cytogenetic assays that can detect radiation-induced DNA damage and misrepair. Blood circulates completely throughout the whole body, so it is reported as the average of the radiation dose received from all parts of the individual body, therefore blood samples should be collected several days or weeks following the exposure to allow complete circulation. If blood sample collection is delayed by several weeks or more, dosimetric results should be reported corrected for the half-life of the biomarker. Newly announced biodosimetry procedures are based on omic approaches. It is impossible to have a single and ideal biodosimetry method in accident dosimetry, where there are many confounding factors and biological factors. Its best management will probably come from the combination of two or more sequential approaches designed for specific accident scenarios.

Important features of biological dosimetry are the dose range and minimum post-exposure time over which the test is informative, the stability of the marker, and the time between the first result and sample collection. Easy and practical use, such as the requirement for specialised technical skills or degree of automation. The invasiveness of sampling required for this dosimetric technique, the yield of analysis and accuracy obtained from the sample. Technical competence to eliminate the influence of confounding factors such as age, gender, other exposures, lifestyle, etc. on the results. Features may include the capability to detect whole-body or partial-body exposures, internal emitters, exposure of low dose rate, exposure of neutron, or the presence of high LET emitter isotopes to obtain more distinctive results (IAEA, 2024).

5.2.1. Dicentric Chromosome Assay (DCA)

During the repair of DNA strand breaks induced by exposure to ionising radiation, chromosomal miss-repair and abnormality in chromosomal replication cause the formation of dicentric chromosomal structure. The technique is considered the "gold standard" for biological dosimetry. One of the disadvantages of DCA is that dicentric chromosomes deviate from the calibration curve at above 5 Gy dose exposure and the exposure dose could be reported lower than the real, and the other is that it needs 2 days of culture and additional time for scoring for output data. Since this assay requires intense effort and experience—both of which are time-consuming—strategies to increase efficiency are important for use in major radiation accidents (Romm et al., 2011).

5.2.2. Premature Chromosome Condensation (PCC)

The PCC assay was introduced to cope with the defects of DCA for high levels of exposure. With this procedure, chromosomes are forced to condense at any time at the cell cycle phase, allowing the detection of individual chromosomes and dicentrics, rings, or translocations for the chromosomal aberrations (Meenakshi et al., 2021). The method is effective over a wide exposure dose range between 0 Gy to 20 Gy, particularly at acute high dose exposures of high and low LET radiation. After blood sampling within a few hours and maintains high accuracy over this dose range (Meenakshi et al., 2021). Like the DCA test, the PCC test is suitable for acute radiation exposure incidents but is restricted in cases of internal exposure to radionuclides, chronic or long-term radiation exposure.

5.2.3. Cytokinesis-Block Micronucleus (CBMN) Cytome Assay

Among *in vivo* and *in vitro* cytogenetic methods, the cytokinesis-block micronucleus (CBMN) assay is one of the most important after DCA (Sioen et al., 2020). Micronuclei (MNs) arise from the generation of small fragments of chromosome substance that fail to interact during anaphase with the mitotic spindle and eventually in the cytoplasm form small, isolated nuclei. Ionising radiation, which induces the formation of unrepaired DNA DSBs, can also induce MNs in a dose-dependent manner during cell division, however, MNs are not uniquely induced as a result of exposure to ionising radiation. The CBMN assay is referred to by counting MNs only in binucleated cells. CBMN testing is faster and requires less expertise than many cytogenetic methods (Sioen et al., 2020). With modifications in CBMN evaluation, the dose range for triage bio dosimetry can be expanded up to 6 Gy (Beinke et al., 2016).

5.2.4. Translocation Analysis by FISH

During the application of the DCA and the CBMN cytome techniques induced damage to be measured is unstable and vanishes sometime after exposure. However, translocations, even if not all are stable, exist for longer periods in lymphocyte structures and after a long time could be utilised to measure exposed doses. Translocation analysis has become a suitable method for biodosimetry with the development of fluorescent, whole-chromosomal staining. FISH: longer and more complex protocol for staining and higher cost. Limited use in emergency biodosimetry because it is an after-a-long time procedure. It can be used in radiation dose exposures greater than 1 Gy. The formation of translocations depends on exposure to various genotoxic agents as well as radiation exposure. Therefore, it is appropriate to report only the translocation frequency followed by a comparison with the background yield by age class (Grégoire et al., 2018).

5.2.5. The γ H 2 AX Assay

Radiation exposure-induced DSBs are the most biologically prominent lesion. Early primary cellular responses to DSBs in the DNA is the serine 139 of histone H2AX phosphorylation, which forms foci in one-to-one correspondence at the DSB region. Visualisation of foci in DSB regions is possible, by this phosphorylated form of H2AX (γ H 2 AX) with used fluorescent markers. In lymphocytes, the dose response is linear up to 10 Gy and doses low a few mGy can be measured under well-controlled conditions. It is easy to process and analyse samples immediately. Between individuals γH2AX background levels vary and are affected by factors such as co-morbidity, age, smoking and lifestyle (IAEA, 2024).

5.2.6. Emerging Assays: Omics

Complex cascades of signalling events triggered by radiation exposure induce multiple changes in the cellular transcriptional process. Also changes in the localisation, activation states, and levels of both effector and signal transduction proteins. As a result of these major changes occur in the metabolism of cells, tissues, and organisms. To quantify changes in the levels of RNA, levels of various proteins, and levels of small molecule metabolites biodosimetry assays are developed. To assess transcriptomics, proteomics, and metabolomics, "omics" technologies improved. The "omics" approaches are not standardised yet and are still in progress (IAEA, 2024).

5.3. Dose Reconstruction

The selection of the dosimetric method depends on the type of exposed radiation, the number of affected individuals, and also on the aim of the measurement. In Table 3 some retrospective dosimetry methods and their features are listed. Some of these are only for short-term dose evaluation, DCA, CBMN, PCC, OSL, and TL. Others can be utilised either in a short time or in a long time after the accident or incident. FISH and EPR are more stable. Remember for the dosimeters that a physical one and a biological one will not produce the identical dose result. If the dosimetric purpose is to assist health experts while determining the most suitable interventional treatment or procedure by performing rapid screening for triage, more detailed monitoring is required. Current emergency dosimetry methods can estimate the original absorbed exposure dose with an uncertainty range of ±20%. According to laboratory comparison results, the OSL of electronic components is probably the one that best reflects the actual situation. The electron paramagnetic resonance technique for biologically derived material is advanced for in-vitro assay but still not sufficient for in-vivo assay (IAEA, 2024).

The dose is reconstructed using combinations of techniques, modelling, simulations, calculations and measurements developed to estimate the exposed radiation dose. Dose reconstruction long after exposure depends on more particular descriptions of the radiation exposure scenario using various sorts of obtained data. And previously developed and benchmarked computer simulation codes, and calculation tools that can be applied quickly and instantly, are the basis for dose reconstruction shortly after exposure (IAEA, 2024).

Physical or biodosimetry and numerical approaches are often combined to normalise the relative dose distribution inside the body to the measured absorbed dose. Digital phantoms with a high degree of anatomical complexity to mimic reality are very effective when simulating real conditions. As an experimental approach, if accident conditions can be safely reproduced, the absorbed dose distribution inside the body can be estimated using a tissue-equivalent phantom with inserted convenient dosimeters (IAEA, 2024).

Table 3. Retrospective physical and biological dosimetric methods and characteristics. IR: Irradiation, BgS: Background Signal, UV: Ultraviolet, X: X rays, γ: Gamma rays β: Beta rays

Assay type	Duration of use after exposure	Identifies partial or whole-body exposure	Time period between sample collection and dose reading	Specificity of Radiation	Dose Exposure Range
Surface mounted components for TL or OSL	Days to weeks	No	1-2 h	X, γ, β	0,1-10 Gy
Integrated circuits for TL or OSL	Days to weeks	No	1-2 h	X, γ, β	0,01-10 Gy
Phone glass for TL or PTTL	Days to weeks	No	1-4 h	BgS, X, γ, UV,	0,3-20 Gy
Other electronic parts for OSL	Days to weeks	No	1-2 h	BgS, X, γ	0,01-10 Gy
Clothing fabrics for OSL	Days to weeks	No	1 h	BgS, X, γ, β	0,1-10 Gy
Other samples for TL or OSL	Days to weeks	No	1 h	BgS, X, γ, β	0,1-10 Gy
Dental materials for OSL	Days to weeks	No	1 h	BgS, X, γ, β, UV	0,01-10 Gy
Teeth for EPR	Days to years	Possible	1 h	BgS, X, γ, β, UV	0,01-10 Gy
Bone for EPR	Days to weeks	Possible	Several hours	X, γ, BgS	1-10 Gy

Table 3. (Continued)

Assay type	Duration of use after exposure	Identifies partial or whole-body exposure	Time period between sample collection and dose reading	Specificity of Radiation	Dose Exposure Range
Nails for EPR	Days	Possible	2-4 h	BgS, X, γ, β, UV	0,1-10 Gy
Phone glass for EPR	Days to weeks	No	1 h	BgS, X, γ, β, UV	>1 Gy
Other samples for EPR	Days to weeks	No	1 h	BgS, X, γ, β, UV	>1 Gy
DCA	Days to months	Yes	55 h	IR	0,1-5 Gy
PCC fragments	Days	Yes	2 h	IR	0,2-20 Gy
Micronuclei	Days to months	No	75 h	BgS, IR	0,2-4 Gy
FISH	Days to years	No	120 h	BgS, IR	0,25-4 Gy
Gamma H2AX	Hours	Yes	3 h	BgS, IR	0,5-10 Gy
Gene expression	Hours to days	Yes	4 h	BgS, IR	0,5-10 Gy
Small metabolits	Hours to days	Yes	3 h	BgS, IR	1-10 Gy
Proteomics	Hours to days	Yes	3 h	BgS, IR	0,5-10 Gy

A combination of computational and retrospective approaches is an option for dose reconstruction. Biological and physical dosimetry with Monte Carlo calculations can be combined to evaluate the dose applied to the organs of interest. Voxel phantoms created by detailing the individual's images obtained by computerised tomography and magnetic resonance imaging can be used (IAEA, 2024).

Future developments, such as 3-dimensional representation of the accident environment or 4-dimensional simulations that take into account the movement of the individual at the time of exposure and after, could quickly create a precise data file for Monte Carlo algorithm (IAEA, 2024).

5.4. Accident Dosimetry for Neutrons

An unplanned and uncontrolled nuclear excursion is the result of the accidental accumulation of some fissile material in a physical geometry whose configuration could support a chain reaction" can lead to neutron criticality accidents. Neutron chain reactions begin as a result of the accumulation of a critical mass or volume of fissile material, involving only fission material or fission material close to hydrogenous material immersed in water. This can cause an immediate accident with a very high level of neutron dose, requiring individual monitoring with a special criticality dosimeter. Specific types of dosimeters are needed instead of routine neutron dosimeters (IAEA, 2024).

Neutron-matter interactions can raise the levels of secondary particles with high LET and major biological effects. Particular methods and additional approaches like measurement of activated radioactivity have to be considered for emergency dosimetry when dealing with neutrons. The biological activities of neutrons differ from gamma rays. Determination of the neutron and the gamma-ray dose components separately is important. As a result of thermal neutron interactions in the body gamma rays are generated. Direct gamma dose element disturbing the body either comes from the neutron production progress or neutron scattering outside the body (IAEA, 2024).

For criticality neutron dosimeters IAEA makes some recommendations. The first exposed dose ranges between 0,25 Gy to 10 Gy, and dose rates up to 10^{5} Gy/s. Then immediate separation of exposed and non-exposed individuals and estimation of the neutron energy spectrum. And reconstruction of the dose within 2 days with < 50% uncertainty or reconstruction of dose within 6 days with < 25% uncertainty. Traditionally, the most common critical accidental dosimeters are activation foils. Detectors for this technique are chosen based on their cross-section and half-live features. A half-life of >1 hour is long enough to let recover and count after the accident. Cross sections are enough to cover thermal neutron energy to > 10 MeV. It is not sensitive to gamma rays so it can detect them individually (IAEA, 2024).

Low-energy Thermal neutrons generally use n, γ reactions and cause a relatively insignificant fraction of the tissue dose. Utilises foil pairs and covers w/ and w/o Cd (Zhao et al., 2021). Thermal neutron fluxes differ significantly from those at the uniform point in the air, varying significantly near the surface of the object. Because the neutrons entering the body are scattered back with the same or lower energies and come out again. Approximately 80% of incident neutrons are backscattered neutrons (IAEA, 2024).

Detectors for intermediate energy neutrons are generally Cd-covered n, γ detectors, and they should have a high sensitivity in the range of 10,2 - 10,5 eV. Of course, that is not an easy achievement, so info on this energy range is often missing. Many factors, such as the angle of incidence, neutron spectrum, and separation of the body-detector, affect the correlation between incident and reflected fluences (IAEA, 2024).

A series of factors impact the choice of nuclear reactions for the dosimetric measurement of fast neutrons. These factors include a low neutron energy threshold for activation detectors, a well-defined cross-sectional area, a properly long half-life, high fast neutron sensitivity, and no attempt for reaction with thermal neutrons that masks the measured reaction product (Tawade et al., 2024). Still, there is no fast neutron detector that satisfies all necessities. Mostly the application of a combined few activation foil is better.

At the time of the accident, individuals do not always wear a dosimeter. Even if they wear a dosimeter, dosimetry for additional dose information would be useful. Accidents accompanied by a neutron component cause the induction of radioactivity in exposed objects.

Neutron activation can be detected inside the body through sodium activation (Na) in a blood sample and can guide the triage. $^{23}Na(n, \gamma)\ ^{24}Na$, t ½ = 14.96 h, E γ 100% is 1.36 MeV and 99.85% is 2.75 MeV.24 Na-induced activity is primarily based on thermal neutrons (Zhao et al., 2021). Comparison of fast neutrons with thermal neutrons, thermal neutrons will produce a lower level of activated Na, however a much higher level of dose. The total activity of Na depends on the exposed area surface of the body and also depends on the exposed mass of the body (IAEA, 2024).

Another option, activated sulfur (S) exists in hair, nails, and wool and could be examined. $^{32}S\ (n,p)\ ^{32}P$, t ½= 14.28 d, $E_{\beta\,max}$ 1.710 MeV (100%). The cross-section of this reaction is almost constant between 3 MeV and 20 MeV and the threshold of the reaction is approximately 3 MeV. Sulfur activity is almost proportional to the neutron kerma in biological samples. From different parts of the body it is easy to collect hair, where it is appropriate to compile information to reflect body dose distribution (IAEA, 2024).

5.5. Contamination

In a radiation accident, internal and external contamination is likely for the public, the occupationally exposed workers, and the emergency workers. It is necessary to know the pattern of contamination to choose the best test type and instrument to use in dosimetry. The level of contamination changes with the presence of internal contamination, the physical properties of the contaminated part of the body, and the chemical or physical form of the contaminants. Evaluating the type and level of contamination is vital for deciding medical strategies. Comprehensive procedures for contaminated workers are already planned. However, contamination of normal people is much harder to manage. Monitoring of the population is a need to evaluate, the required medical treatment, the presence of radioactive materials in the body (internal contamination), the presence of radioactive contamination on the body or clothing (external contamination), the radiation dose received, and the resulting health risks from the exposure.

Evaluation of internal contamination is more difficult than evaluation of external contamination. For external decontamination, hair can be cut or washed, clothing can be removed, and skin can be treated with various decontamination agents in physical and chemical forms. On the other hand, it is difficult to give numerical the degree of external contamination of the body, cause maybe it is contaminated very heterogeneously. Several types of devices are acceptable for contamination detection of gamma, alpha, and/or beta radiation emitters. Some of these include measuring devices for surface contamination of alpha and beta, portable monitors, gamma dose rate devices and portable gamma spectrometers, and personal alarm dosimeters that can measure the cumulative dose and instantaneous dose rate (IAEA, 2024).

Skin contamination, a serious condition that must be managed, will cause external exposure, and depending on the type of radionuclide, the chemical form of radionuclide, and the activity concentration of the radionuclide sometimes cause internal exposure. The main aim is to maintain the prevention of deterministic effects and, in case of overexposure, to pioneer appropriate medical examinations and interventions (IAEA, 2024).

6. What About the Inverse Exposure Estimation Based on Acute Radiation Syndrome Symptoms?

This part is different from an actual dose measurement technique, but it is a result of various very valuable analyses. Various guidelines have been published for the triage and management of Acute Radiation Syndrome abbreviated as ARS, which develops as a result of radiation accidents. A multinational and multidisciplinary guideline for ARS, specifically for radiation-related emergencies, was first published in 1997 by the Commission of the European Communities under the name Medical Treatment Protocols for Radiation Accident Victims as a Basis for a Computerised Guidance System (METREPOL) (López and Martín, 2011).

Is it possible to make accidental dose estimations based on the ARS symptoms? Not every incident or accident scenario occurs in radiation areas with radiation recording equipment or a warning signal that detects the presence of radiation. In some accident reports, it has been declared that individuals consulted a hospital or medical centre with a few symptoms days after exposure. If the patient gives information about possible radiation exposure, while waiting for the test results related to the patient's symptoms, it is also necessary to triage the patient and make some interventions.

Acute radiation syndrome is a high-risk, low-frequency condition that can be fatal and difficult to diagnose (Ebeling et al., 2023). Acute radiation syndrome is a dose-related set of symptoms that manifest themselves after short-term exposure to high levels of ionising radiation (Christensen, Iddins and Sugarman, 2014).

Classically, it has been stated that for ARS to develop, whole-body or partial-body exposure must be at least ≥ 1 Gy, and ARS is not expected for exposures below 0.5 Gy (López and Martín, 2011). Symptoms are divided into four subsyndromes based on the observed system and dose threshold: ≥ 1 Gy is hematopoietic, ≥ 3 Gy is cutaneous, ≥ 6 Gy is gastrointestinal, and ≥ 8 Gy is neurovascular (Table 4) (Dainiak et al., 2003).

In hematopoietic subsyndrome (≥ 1 Gy), ionising radiation damages radiosensitive lymphocytes and stem cells inside the bone marrow, pancytopenia develops, and it becomes easier for opportunistic infections to enter the body (Christensen, Iddins, Parrillo, et al., 2014). A 50% decrease in absolute lymphocyte count within the first 24 hours after exposure, but a more severe decrease in the following 48 hours, indicates a potentially lethal higher exposure between 5 to 10 Gy. Lymphocyte nadir develops 8-30 days after exposure (Goans et al., 2001).

In cutaneous subsyndrome (≥ 3 Gy), ionising radiation exposure can induce epilation, erythema, exfoliation, and even radionecrosis because the skin comprises variable and rapidly dividing cells, as do the gastrointestinal tract and the bone marrow (Iddins et al., 2022). The cutaneous syndrome may develop early following exposure (e.g., 1–2 days) (López and Martín, 2011).

In gastrointestinal (GI) subsyndrome (≥ 6 Gy), ionising radiation-induced injury in the intestine can cause a wide diversity of problems, including bleeding, electrolyte imbalance, and severe infection. Symptoms appear within 5 days following exposure. Radiation exposure dose ≥ 1.5 Gy: prodromal symptoms: nausea, vomiting and diarrhoea. Radiation dose between 5-12 Gy: the destruction of intestinal crypt cells, impaired mucosal barrier, decreased nutrient absorption: abdominal pain, vomiting, GI bleeding. Dehydration, electrolyte imbalances, high fever and diarrhoea due to bacterial translocation, intestinal wall necrosis and subsequent perforation symptoms, which can lead to death (Lombardini et al., 2013).

Finally, the neurovascular subsyndrome (≥ 8 Gy) localised central nervous system changes due to the whole body exposure dose excess of 10 Gy. Loss of consciousness, fever, hypotension, and cerebral oedema are all associated with neurovascular subsyndrome (López and Martín, 2011).

Local changes in the cerebrovascular system cause impaired capillary circulation, damage to the blood-brain barrier, hyperexcitability response, interstitial oedema, acute inflammation in the meninges and petechial haemorrhages. Headaches, nausea, mental status changes or seizures may be the main complaints of individuals. Physical examination results could be observed as papilledema, ataxia, decreased or absent deep tendon reflexes and corneal reflexes (Theodor, Fliedner and Friesecke, 2001). Regardless of subsyndromes, prodromal, latent, manifest disease, and recovery/death are the four stages that ARS goes through. The higher the exposed dose, the faster the exposed patient progresses through these stages. Prodromal Phase: Begins 0-2 days after exposure, Latent Phase: Begins 2-20 days after exposure, Manifest Disease: Begins 21-60 days after exposure, see Table 5 (ORISE, 2017).

Some threshold values have been determined for the lethal dose (LD) LD 50/60 exposure dose limitations that would kill 50% of the exposed community within 60 days, Table 6. Without supportive care, LD 50/60 is 3,5-4,0 Gy, with antibiotics or supportive care, LD 50/60 is 5,0-6,0 Gy, and with easy access to intensive care, LD 50/60 may increase to 7,0-9,0 Gy (ORISE, 2017).

Table 4. Acute Radiation Subsyndromes – Dose - Prodromal Phase Indicators

	Nausea, Vomiting and Diarrhoea NVD Subsyndrome	Hematopoietic Subsyndrome	Gastrointestinal Subsyndrome	Neurovascular Subsyndrome
The Magnitude of Absorbed Dose	1-2 Gy	2-6 Gy	5-12 Gy	8-20 Gy
Prodromal Phase Indicator Symptoms	Nausea, Vomiting, Diarrhoea, Anorexia, Loss of petit.	Loss of cellularity of bone marrow, spleen, and thymus	Damage to intestinal crypt cells, loss of mucosal barrier, loss of weight, dehydration, low blood pressure	Irritability, confusion, burning sensations of the skin, epileptic type fits, coma.

Table 5. Acute radiation syndrome phase - time correlation

Days	Phase
Within 0-2 days following exposure	Prodromal
Within 2-20 days following exposure	Latent
Within 21-60 days following exposure	Manifest Disease
60 days and more following exposure	Recovery Or Death

Table 6. Exposure dose - Acute radiation syndrome - LD 50/60 correlation

Dose	1 Gy	Hematopoietic syndrome	(0–2 Gy) 100% survival without treatment
	6 Gy	GastroIntestinal syndrome	(6–7 Gy) LD50/60 with supportive care (3–5 Gy) LD50/60 without treatment
	≥ 20 Gy	Neurovascular syndrome	(>10 Gy) Multiple organ failure probable death (8–10 Gy) Consider stem cell transplant

It has been argued that time and resources should not be spent on definitive treatment of whole-body exposures over 10 Gy, especially in mass events, and that palliative care will be sufficient because it is unlikely that these patients will survive (Christensen et al., 2014). Unfortunately, ARS has been diagnosed late even if it manifests itself (Koenig et al., 2005).

Some guidelines have been published as a result of the difficulty in diagnosing ARS due to its nonspecific symptomatology for an individual without a clear history of exposure or accident. The METREPOL guideline defines the first step of treatment as taking a good history regarding the type, duration and dose of exposure, removing external contamination and fluid-electrolyte replacement. It has been reported that people exposed to a dose of 2-3 Gy must be isolated due to the risk of infection (Theodor, Fliedner and Friesecke, 2001).

Early symptoms that manifest themselves in the prodromal phase include nausea, vomiting, anorexia, apathy, diarrhoea, fever, headache, and tachycardia. If these symptoms started within about 2 hours, the guideline stated that the patient was most likely exposed to a high dose, or if they started within minutes, then the patient was exposed to doses exceeding 10 Gy (López and Martín, 2011).

Conclusion

In appropriate dose measurement techniques and calculations that evaluate the real conditions of the incident or accident are very important. However, the ARS symptomatic data gives us a lot of information about both estimations of absorbed dose corresponding to the exposure dose of individuals and through observations on severity and manifestation period of symptoms corresponding to dose rate. The correlation of ARS symptoms and dosimetric data of exposure and dose rate shows us that we are at the right point when verifying the calculations derived from the site of the accident.

Exposure and dose rate measurements in radiation accidents are vital procedures to be done and well documented. Since the first radiation accident was reported up to the present day a huge knowledge has formed about dosimetry and hazard estimations with data obtained from major radiation accidents and case reports around the world. Therefore it is very valuable to declare and document the experiences, incidents and accidents with all its details and estimations.

References

Barazzuol, L., Coppes, R. P., & van Luijk, P. (2020). Prevention and treatment of radiotherapy-induced side effects. *Molecular Oncology*, *14*(7), 1538–1554. https://doi.org/10.1002/1878-0261.12750.

Bassinet, C., Trompier, F., & Clairand, I. (2010). Radiation Accident Dosimetry On Electronic Components By Osl. *Health Physics*, *98*(2), 440–445. https://doi.org/10.1097/01.HP.0000346335.56701.93.

Beinke, C., Port, M., Riecke, A., Ruf, C. G., & Abend, M. (2016). Adaption of the Cytokinesis-Block Micronucleus Cytome Assay for Improved Triage Biodosimetry. *Radiation Research*, *185*(5), 461. https://doi.org/10.1667/RR14294.1.

Bentzen, S. M. (2006). Preventing or reducing late side effects of radiation therapy: radiobiology meets molecular pathology. *Nature Reviews Cancer*, *6*(9), 702–713. https://doi.org/10.1038/nrc1950.

Cameron, J. (1991). Radiation dosimetry. *Environmental Health Perspectives*, *91*, 45–48. https://doi.org/10.1289/ehp.919145.

Cerezo, L. (2011). Radiation accidents and incidents. What do we know about the medical management of acute radiation syndrome? *Reports of Practical Oncology & Radiotherapy*, *16*(4), 119–122. https://doi.org/10.1016/j.rpor.2011.06.002.

Christensen, D. M., Iddins, C. J., Parrillo, S. J., Glassman, E. S., & Goans, R. E. (2014). Management of Ionizing Radiation Injuries and Illnesses, Part 4: Acute Radiation Syndrome. *Journal of Osteopathic Medicine*, *114*(9), 702–711. https://doi.org/10.7556/jaoa.2014.138.

Christensen, D. M., Iddins, C. J., & Sugarman, S. L. (2014). Ionizing Radiation Injuries and Illnesses. *Emergency Medicine Clinics of North America*, *32*(1), 245–265. https://doi.org/10.1016/j.emc.2013.10.002.

Christensen, D. M., Jenkins, M. S., Sugarman, S. L., & Glassman, E. S. (2014). Management of Ionizing Radiation Injuries and Illnesses, Part 1: Physics, Radiation Protection, and Radiation Instrumentation. *Journal of Osteopathic Medicine*, *114*(3), 189–199. https://doi.org/10.7556/jaoa.2014.037.

Dainiak, N., Waselenko, J. K., Armitage, J. O., MacVittie, T. J., & Farese, A. M. (2003). The Hematologist and Radiation Casualties. *Hematology*, *2003*(1), 473–496. https://doi.org/10.1182/asheducation-2003.1.473.

Ebeling, M., Bloom, A., Boggiano, M. M., Peterson, D. T., & Peterson, T. (2023). Employing High-Fidelity Simulation for the High-Risk, Low-Frequency Diagnosis and Management of Acute Radiation Syndrome (ARS). *MedEdPORTAL : The Journal of Teaching and Learning Resources*, *19*, 11331. https://doi.org/10.15766/mep_2374-8265.11331.

Galindo, A. (2023). *What is Radiation?* https://www.iaea.org/newscenter/news/what-is-radiation.

George A. Alexander. (2016). *Ciottone's Disaster Medicine* (Gregory R. Ciottone (Ed.)).

Goans, R. E., Holloway, E. C., Berger, M. E., & Ricks, R. C. (2001). Early Dose Assessment In Criticality Accidents. *Health Physics*, *81*(4), 446–449. https://doi.org/10.1097/00004032-200110000-00009.

Goodarzi, A. A., Anikin, A., & Pearson, D. D. (2016). Environmental Sources of Ionizing Radiation and Their Health Consequences. In *Genome Stability* (pp. 569–581). Elsevier. https://doi.org/10.1016/B978-0-12-803309-8.00033-1.

Grégoire, E., Roy, L., Buard, V., Delbos, M., Durand, V., Martin-Bodiot, C., Voisin, P., Sorokine-Durm, I., Vaurijoux, A., Voisin, P., Baldeyron, C., & Barquinero, J. F. (2018). Twenty years of FISH-based translocation analysis for retrospective ionizing radiation biodosimetry. *International Journal of Radiation Biology*, *94*(3), 248–258. https://doi.org/10.1080/09553002.2018.1427903.

Harshman, A., & Johnson, T. (2019). Dose Reconstruction Using Electron Paramagnetic Resonance Dosimetry on Tooth Enamel From Wild Boar Living in the Fukushima Exclusion Zone. *Health Physics*, *116*(6), 799–806. https://doi.org/10.1097/HP.0000000000001040.

IAEA. (2024). *Assessment of Occupational Exposure due to External Radiation Sources.* Assessment of Occupational Exposure due to External Radiation Sources.

Iddins, C. J., DiCarlo, A. L., Ervin, M. D., Herrera-Reyes, E., & Goans, R. E. (2022). Cutaneous and local radiation injuries. *Journal of Radiological Protection*, *42*(1), 011001. https://doi.org/10.1088/1361-6498/ac241a.

International Atomic Energy Agency [IAEA]. (2024). *IAEA*. www.iaea.org.

International Commission on Radiological Protection [ICRP]. (2007). *The 2007 Recommendations of the International Commission on Radiological Protection. ICRP Publication 103. Ann. ICRP 37 (2-4).*

Koenig, K. L., Goans, R. E., Hatchett, R. J., Mettler, F. A., Schumacher, T. A., Noji, E. K., & Jarrett, D. G. (2005). Medical Treatment of Radiological Casualties: Current Concepts. *Annals of Emergency Medicine*, *45*(6), 643–652. https://doi.org/10.1016/j.annemergmed.2005.01.020.

LiVolsi, V. A., Abrosimov, A. A., Bogdanova, T., Fadda, G., Hunt, J. L., Ito, M., Rosai, J., Thomas, G. A., & Williams, E. D. (2011). The Chernobyl Thyroid Cancer Experience: Pathology. *Clinical Oncology*, *23*(4), 261–267. https://doi.org/10.1016/j.clon.
2011.01.160.

Lombardini, E. D., Pacheco-Thompson, M. E., & Melanson, M. A. (2013). Radiation and Other Physical Agents. In *Haschek and Rousseaux's Handbook of Toxicologic Pathology* (pp. 1421–1503). Elsevier. https://doi.org/10.1016/B978-0-12-415759-0.00044-3.

López, M., & Martín, M. (2011). Medical management of the acute radiation syndrome. *Reports of Practical Oncology & Radiotherapy*, *16*(4), 138–146. https://doi.org/10.1016/j.rpor.2011.05.001.

Lowe, D., Roy, L., Tabocchini, M. A., Rühm, W., Wakeford, R., Woloschak, G. E., & Laurier, D. (2022). Radiation dose rate effects: what is new and what is needed? *Radiation and Environmental Biophysics*, *61*(4), 507–543. https://doi.org/10.1007/s00411-022-00996-0.

Maltar-Strmečki, N., Bortolin, E., Kenzhina, L., Patrono, C., & Testa, A. (2022). Editorial: Developing novel materials and new techniques of biological and physical retrospective dosimetry for affected individuals in radiological and nuclear emergencies. *Frontiers in Public Health*, *10*, 1117269. https://doi.org/10.3389/fpubh.2022.1117269.

Meenakshi, C., Venkatachalam, P., Chandrasekaran, S., & Venkatraman, B. (2021). Construction of dose response curve for 6 MV LINAC X-rays using Premature Chromosome Condensation assay for radiation dosimetry. *Applied Radiation and Isotopes*, *173*, 109729. https://doi.org/10.1016/j.apradiso.2021.109729.

Oak Ridge Institute for Science and Education [ORISE]. (2017). *The Medical Aspects Of Radiation Incidents*. https://orise.orau.gov/reacts/index.html.

Percuoco, R. (2014). Plain Radiographic Imaging. In *Clinical Imaging* (pp. 1–43). Elsevier. https://doi.org/10.1016/B978-0-323-08495-6.00001-4.

Reeves, G. I. (1999). Radiation Injuries. *Critical Care Clinics*, *15*(2), 457–473. https://doi.org/10.1016/S0749-0704(05)70063-4.

Romm, H., Wilkins, R. C., Coleman, C. N., Lillis-Hearne, P. K., Pellmar, T. C., Livingston, G. K., Awa, A. A., Jenkins, M. S., Yoshida, M. A., Oestreicher, U., & Prasanna, P. G. S. (2011). Biological Dosimetry by the Triage Dicentric Chromosome Assay: Potential Implications for Treatment of Acute Radiation Syndrome in Radiological Mass Casualties. *Radiation Research*, *175*(3), 397–404. https://doi.org/10.1667/RR2321.1.

Rühm, W., Azizova, T., Bouffler, S., Cullings, H. M., Grosche, B., Little, M. P., Shore, R. S., Walsh, L., & Woloschak, G. E. (2018). Typical doses and dose rates in studies pertinent to radiation risk inference at low doses and low dose rates. *Journal of Radiation Research*, *59*(suppl_2), ii1–ii10. https://doi.org/10.1093/jrr/rrx093.

Rühm, W., Woloschak, G. E., Shore, R. E., Azizova, T. V., Grosche, B., Niwa, O., Akiba, S., Ono, T., Suzuki, K., Iwasaki, T., Ban, N., Kai, M., Clement, C. H., Bouffler, S., Toma, H., & Hamada, N. (2015). Dose and dose-rate effects of ionizing radiation: a discussion in the light of radiological protection. *Radiation and Environmental Biophysics*, *54*(4), 379–401. https://doi.org/10.1007/s00411-015-0613-6.

Shubayr, N., & Alashban, Y. (2021). Assessment of Hp(10) and Hp(0.07) doses for cardiac catheterization personnel: A 5-year retrospective study. *Radiation Physics and Chemistry*, *185*, 109517. https://doi.org/10.1016/j.radphyschem.2021.109517.

Sioen, S., Cloet, K., Vral, A., & Baeyens, A. (2020). The Cytokinesis-Block Micronucleus Assay on Human Isolated Fresh and Cryopreserved Peripheral Blood Mononuclear Cells. *Journal of Personalized Medicine*, *10*(3), 125. https://doi.org/10.3390/ jpm10030125.

Tanaka, T., Matsubara, K., Fukuda, A., & Kobayashi, S. (2019). Estimation Of Hp(3) To The Eye Lens Of Interventional Radiologists—Relation Between The Eye Lens Dose And Radiologist's Height.*Radiation Protection Dosimetry*, *187*(4), 409–417. https://doi.org/10.1093/rpd/ncz181.

Tawade, N. S., Tripathi, R., Nag, T. N., Patra, S., Datrik, C. S., Pujari, P. K., Thomas, R. G., Mishra, G., Kumar, A., De, S., & Kumawat, H. (2024). Measurement of fast neutron induced (n,γ) reaction cross-section of 152Sm, 154Sm and 150Nd in the energy range of 0.8 to 2 MeV. *Nuclear Physics A*, *1042*, 122805. https://doi.org/ 10.1016/j.nuclphysa.2023.122805.

Theodor M., Fliedner, I. Friesecke, K. B. (Ed.). (2001). *Medical Management of Radiation Accidents: Manual on the Acute Radiation Syndrome*. British Institute of Radiology.

Thorne, M. C. (2003). Background radiation: natural and man-made. *Journal of Radiological Protection*, *23*(1), 29–42. https://doi.org/10.1088/0952-4746/23/1/302.

Tuieng, R. J., Cartmell, S. H., Kirwan, C. C., & Sherratt, M. J. (2021). The Effects of Ionising and Non-Ionising Electromagnetic Radiation on Extracellular Matrix Proteins. *Cells*, *10*(11), 3041. https://doi.org/10.3390/cells10113041.

West, C. M. L. (1995). Intrinsic radiosensitivity as a predictor of patient response to radiotherapy. *The British Journal of Radiology*, *68*(812), 827–837. https://doi.org/10.1259/0007-1285-68-812-827.

World Health Organization [WHO]. (2023). *How does radiation effect my health ?*https://www.who.int/news-room/questions-and-answers/item/radiation-and-health?gad_source=1&gclid=EAIaIQobChMIteqdpvG-hQMVqGhBAh1ewgJmEAAYASAAEgJDa_D_BwE.

Zhao, D., Jia, W., Hei, D., Cheng, C., Li, J., Wang, M., & Qiu, M. (2021). Development of a compensation system for thermal neutron measurements based on Cd(Cu)-covered NaI(Tl) detectors. *Applied Radiation and Isotopes*, *176*, 109882. https://doi.org/10.1016/j.apradiso.2021.109882.

Zidouh, I., Arectout, A., Bellahsaouia, M., Elaarabi, D., Chamlal, H., Maroufi, B., Sadeq, Y., Tazi, M., Rodenas, J., Boukhal, H., & Chakir, E. (2023). Comparison of OSL and TL dosimetry systems against IEC and ICRP standards. *Applied Radiation and Isotopes*, *196*, 110732. https://doi.org/10.1016/j.apradiso.2023.110732.

Zoetelief, J., & Broerse, J. J. (1990). Dosimetry for Radiation Accidents: Present Status and Prospects for Biological Dosemeters. *International Journal of Radiation Biology*, *57*(4), 737–750. https://doi.org/10.1080/09553009014550901.

Chapter 12

Major Radiation Accidents in the World and Case Studies

Nadir Küçük*

Anadolu Medical Center, Radiation Oncology Department, Kocaeli, Türkiye

Abstract

Radiation technologies play a critical role in this century. As the use of radiation becomes more widespread in many areas such as medicine and industry, the risks of radiation-related accidents also increase. Due to radiation accidents, people and the natural environment may face some unexpected exposures that may threaten life. In order to minimise the possibilities, it is of great importance to thoroughly investigate the causes of past accidents and reveal the necessary actions and apply them into practice. Only in this way that possible radiation accidents can be prevented in the first place by designing detailed and comprehensive safety measures in every field related to radiation. In this section, major radiation accidents that have occurred in the past are examined and the possible causes of these accidents are detailed through some selected case studies.

Keywords: accident, radiation, unexpected exposure

* Corresponding Author's Email: nadir.kkucuk@gmail.com.

In: Triage in Radiation Accidents
Editors: Nural Öztürk and Sibel Karaca
ISBN: 979-8-89530-415-0

1. Introduction

Radiation is commonly used for various purposes in many different areas in today's technology (Reed, 2011). Although radiation was discovered approximately a century ago, its utilisation in industries, research activities, medicine, and nuclear power plants is increasing day by day and provides significant benefits to humanity. While ionising radiation is commonly used in the diagnosis and treatment of diseases in medicine, it has also become one of the most important tools in the development of medical technologies (Donya et al., 2015). Many diseases threatening public health are treated with different types of radiation used in radiology, nuclear medicine, radiation oncology, and many other different medical areas (Bhargavan et al., 2008). As in many fields of activity, such as medicine, engineering, archaeology, etc., nuclear materials play crucial roles in research, development, and production activities within universities and other academic/scientific institutions (IAEA, 2017). Similarly, another field where radiation is extensively utilised is industry (Raj, 1995; Reisz et al., 2014; National Academies of Sciences, Engineering, and Medicine, 2021). Radiation technologies play a critical role in modern industry. Non-destructive examination and analysis of structural integrity can be easily conducted through radiation (Raj et al., 1995). Radiation sources have the capability to alter various materials used in industry for different commercial applications. Moreover, high-energy radiation can induce chemical and biological effects that render many pollutants less harmful (Reisz et al., 2014). In today's world, one of the most debated yet widely used areas where radiation is concerned is nuclear power plants. Steam, which is used to provide power to generators for electricity production, is typically obtained by boiling water through the process of nuclear fission (Fernández-Arias, 2020; Ho et al., 2019). While nuclear power plants constitute an important source of energy production, they also raise environmental and safety concerns. While the effects of such facilities generally attract the attention of the public and scientists, their potential to meet energy needs is also constantly discussed as an important issue.

As the use of radiation becomes widespread in many areas such as medicine and industry, the risks of radiation-related accidents also increase. If adequate safety precautions are not taken in areas where radiation is used, the damage caused by radiation exposure can be quite severe. Due to radiation accidents, people and the natural environment may face unexpected exposures that can threaten life. To reduce these possibilities, it is of great importance to thoroughly investigate the causes of past accidents and learn the necessary lessons. In this chapter, major radiation accidents that occurred in the past are examined and the possible causes of these accidents are detailed through some selected case studies.

2. Overview of Radiation Accidents

There have been numerous radiation accidents, caused by nuclear reactors, radiation-producing devices, and radioactive materials, over the past 50 years (Nénot, 2009; Saenger, 1986; Coeytaux et al., 2015). A large number of people have been affected by these accidents. These people have been exposed to internal or external radiation, resulting in various permanent damages, including death. As a consequence of these accidents, not only is the overall health of humans damaged, but the surrounding region, including all the living mechanisms, have been dramatically affected.

Radiation has deterministic effects (such as radiation burns, infertility, cataracts, etc.) and stochastic effects (such as cancer, etc.) on living organisms (Mettler, 2012). These effects can manifest immediately or may appear weeks or months later. The importance of these effects depends on several parameters. The total dose, the volume of the body exposed to radiation and the organs it contains, the dose rate, the type of radiation from the source, and the individual characteristics of the persons exposed to radiation (age, performance status, etc.) are considered important parameters that determine the degree of the effects.

Measures are generally taken at the local level in countries to reduce or minimise accident risks in areas where radiation is widely used. Many international organisations are also working on this subject. The United Nations Scientific Committee on the Effects of Atomic Radiation (UNSCEAR), the International Atomic Energy Agency (IAEA) and the International Commission on Radiological Protection (ICRP) are the most important among these organisations. These organisations evaluate radiation risks, make recommendations, and promote the safe use of rapidly evolving and increasingly complex radiation technologies (Carpeggiani et al., 2012). There are many reviews in the literature about radiation accidents prepared by these organisations or other individual volunteers. One of the most important of these reports may be the one prepared by the UNSCEAR (Asano et al., 2000). In 1990, the IAEA and the Organisation for Economic Co-operation and Development developed a scale initially intended to classify incidents at nuclear power plants. However, it was later expanded to meet the growing need for communication regarding the significance of all incidents related to the use, storage, and transportation of radioactive material and radiation sources. This scale, shown in Figure 1, is called The International Nuclear and Radiological Event Scale (INES) (Wheatley et al., 2018).

Preventing radiation accidents and minimising the damage when an accident occurs can only be achieved through an excellent organisation (Barquinero et al., 2021; ESR, 2015). Establishing an organised and trained health physics team, thorough planning, accurately defining dose limits strictly adhering to these rules, and setting action levels are just a few of the important parameters in preventing or minimising accidents.

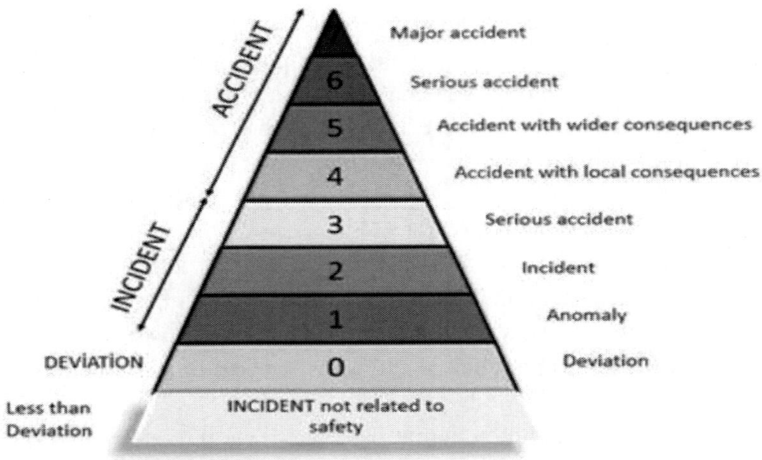

Figure 1. INES tool indicating accident and incident levels developed by the IAEA and the Organisation for Economic Co-operation and Development.

Preparing radiation protection programs, creating emergency response teams, developing methods for communication and intervention with local and national authorities, and providing specific training to the media and the public is crucial for preventing accidents or minimising damage through immediate and organised intervention if accidents occur. Recent nuclear plant accident examples that have occurred worldwide are shown in Table 1 (BBC, 2011; Sovacool, 2010; Sovacool, 2008, Sovacool, 2009). As shown in Table 1, many nuclear reactor accidents with INES of 4 and above have occurred in Asia, Europe and the American continents in the last 70 years. Some of these accidents are discussed in detail in this chapter.

Table 1. Nuclear reactor accidents and their INES ratings over the past 70 years

Date	Location	Description	INES level
Mart-2011	Fukushima-Japan	Trigged by tsunami	7
July-2006	Forsmark-Sweden	Electrical problem	2
August-2004	Fukui Prefecture-Japan	Steam explosion	1
April-2003	Paks-Hungary	Leakage of radioactive gases	3
February-2002	Ohio-USA	Corrosion of reactor vessel head	3
September-1999	Ibaraki Prefecture-Japan	Radiation above limits	4
October-1989	Vandellos-Spain	Fire damage	3
April-1986	Chernobyl-Ukraine	Explosion and meltdown	7
March-1979	Three Mile Island, Pennsylvania-USA	Release of radioactive gases	5
January-1976	Jaslovske Bohunice-Czechoslovakia	A malfunction during fuel replacement	4
December-1975	Greifswald Germany	Fire in the main trough	3
January-1969	Vaud-Switzerland	Loss of coolant accident	4
October-1966	Michigan-USA	Meltdown of fuel elements	4
January-1961	Idaho-USA	Explosion	4
September-1957	Kyshtym-Soviet Union	Explosion	6
October-1957	Cumberland-United Kingdom	Released Iodine-131	5

3. Case Studies

3.1. Fukushima

On March 11, 2011, a major underwater earthquake occurred in the Pacific Ocean near the Oshika Peninsula in the Tohoku region of Japan. The earthquake struck around 14:46 Japan Standard Time, causing massive destruction in the Tohoku region. This earthquake had a magnitude of 9.0 and was one of the largest in Japan's history (Ichiseki, 2013). The earthquake, which lasted approximately six minutes, triggered a tsunami warning. Waves reaching up to 3.6 metres in height were recorded, resulting in one of the largest tsunamis in history (Pletcher, 2024). The tsunami began to expand from its point of origin at approximately 800 km/h and created large waves along the coasts of Hawaii, Shemya Island, and the

California/Oregon coasts in North America. This chain of events resulted in the deaths of approximately 18,500 people, triggered a disaster at the Fukushima Daiichi Nuclear Plant (NPP), and led to the evacuation of 150,000 people from the area (Raskob et al., 2020).

The accident was first reported on March 11, 2011, at 14:56 when the operating reactors automatically shut down due to a power loss (Raskob et al., 2020). Chronology showing hour-by-hour key events on the first day immediately after the Daiichi NPP accident as shown in Figure 2. Immediately following this report, on the first and second day, Fukushima Daiichi NPP reported various types of radiation accidents from their different units with emergency codes to local and national authorities. All 4 reactors at NPP became unusable. Two weeks after the accident, the three reactors (units 1-3) were stabilised by adding water, and by July, they began to be cooled with recycled water from a new treatment facility. Local and national authorities raced against time to immediately begin evacuating the area around the NPP. Initially focusing on the nearby surroundings, they gradually expanded the operation, and by March 16th, they had completely evacuated a 20-kilometre radius. The distribution of evacuated residents from the area surrounding the NPP over the days following the accident could be seen in some published data (Independent, 2011). In December 2011, it was officially announced that the "cold shutdown position" work was completed (World Nuclear Association, 2024). As important as the cooling operations were the efforts to prevent the release of radioactivity. It was crucial to quickly stop the leakage of contaminated water, especially from units 1-3. These efforts continued until they were completed in August 2013.

The tsunami triggered by the earthquake primarily damaged the backup generators at the NPP. Although all three operating reactors were successfully shut down, the power loss caused the cooling systems of each reactor to fail in the first few days of the disaster. The increasing heat in each reactor core led to the fuel rods in reactors 1, 2, and 3 overheating and partially melting, resulting in the release of radiation.

The melted material fell into the containment vessels in reactors 1 and 2, causing large holes to open at the bottom of the vessel. These holes caused the nuclear material in the nuclei to be partially released. Explosions occurred in the outer containment buildings surrounding reactors 1 and 3 due to the accumulation of pressurized hydrogen gas. NPP workers attempted to cool and stabilise cores by pumping seawater and boric acid into the units. The third explosion occurred on March 15th in the building surrounding reactor 2. At the time, it was believed that the explosion had damaged the containment vessel housing the fuel rods. It was later understood that the explosion had created a second hole in the containment vessel, with the first hole having been caused earlier by molten nuclear material passing through the vessel's base. The explosion, along with a fire caused by rising temperatures in the spent fuel rods stored in reactor 4, led to the release of higher levels of radiation from the facility (Pletcher, 2024).

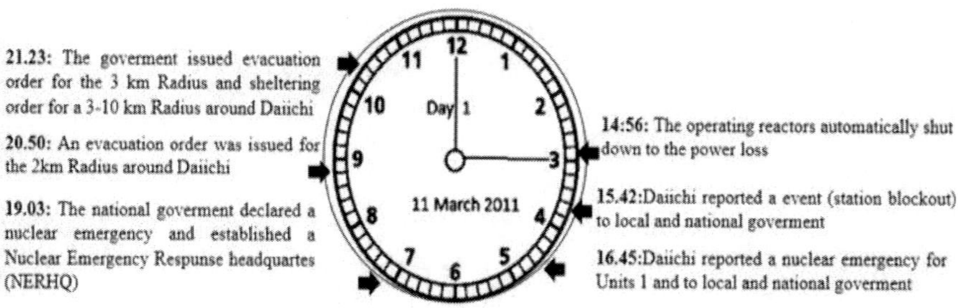

Figure 2. Chronology showing hour-by-hour key events on the first day immediately after the Daiichi NPP accident.

The first group to be exposed to radiation due to the accident was the workers at the NPP. Doses to workers have been reported by Tokyo Electric and Power Company (TEPCO) and some of its contractors. According to TEPCO records, the average effective dose of workers for 19 months after the accident was measured as 12 millisievert (mSv). Of the workers, 0.7% received doses exceeding 100 mSv. According to the UNSCEAR report (UNCEAR, 2013), 12 of these highly exposed workers received thyroid doses ranging from 2-12 Gy due to inhalation of radioactive sources. The maximum and average effective doses received by NPP workers over the years are shown in Table 2. In 2021, UNSCEAR reported that since March 2013, none of the NPP workers had an annual average dose exceeding 50 mSv (UNCEAR, 2022). No deaths or cases of radiation sickness were reported due to the nuclear accident, but as a precautionary measure, over 100,000 people were evacuated from their homes. According to the report on the Great East Japan Earthquake published by the Japan Reconstruction Agency (2012), as of March 31, 2013, there have been 2,688 disaster-related deaths (DRD) (Hasegawa et al., 2016). DRD is generally defined as death resulting from the worsening of underlying medical conditions due to poor access to medical care or diseases caused by poor living conditions, such as temporary shelters, following a disaster. 90% of these cases were aged 66 and above, and more than one-third died within one month after the earthquake. Among the three Tohoku prefectures, Fukushima had the highest number of deaths with 1914 cases, and the national report clearly indicated that the effects of the nuclear accident could be the main reason for the increased mortality rate among the elderly displaced in Fukushima. These findings are supported by Table 2, which illustrates the dramatic decrease in radiation levels in the Fukushima region between April March 2011 and March 2020.

In studies conducted up to the present day, although cytogenetic, physiological, and morphological effects have been observed in some plants and animals in areas where increased radiation levels have been detected after accidents, there are no widely reported studies on group effects. However, the UNSCEAR committee continues to consider that there may not be an impact on regional wildlife populations but harmful effects on individual organisms could still be possible. The studies supporting this view of the committee are still quite limited, and current results strongly depend on the organism under investigation. Several studies have indicated significant population effects on specific wildlife groups, but these studies have considerable uncertainties, and future research needs more focused studies to reduce uncertainties.

3.2. Chernobyl

In the Chernobyl NPP, located near the city of Pripyat, which was then part of the Soviet Socialist Republics Union and approximately 100 km away from the city of Kiev in Ukraine, an accident occurred on April 26, 1986, due to a major explosion. The fire that broke out after the explosion lasted for 10 days, during which an unprecedented release of gas occurred in history (Alexakhin et al., 2006).

The Chernobyl accident took its place in history as the world's largest uncontrolled radioactive release accident. Like the Fukushima Daiichi NPP accident, it has an INES value of 7 and is recorded as one of the greatest environmental catastrophes. The explosion and fire caused hot particles of nuclear fuel along with radioactive isotopes of Cesium-137, Iodine-131, Strontium-90, and other radionuclides to be released into the air. Many of these exposed materials released during the accident accumulated as debris

Table 2. A graph showing the dose levels received by Fukushima Daiichi NPP workers up to the year 2020

Timeline	Average Effective Dose(mSv)	Maximum Effective Dose (mSv)	Number of Worker
March 2011-March 2012	13,0	679	21.135
April 2012-March 2013	5,7	54	13.742
April 2013-March 2014	5,3	42	14.746
April 2014-March 2015	5,0	40	20.730
April 2015-March 2016	4,3	43	18.196
April 2016-March 2017	2,9	39	15.835
April 2017-March 2018	2,7	33	13.943
April 2018-March 2019	2,4	20	11.306
April 2019-March 2020	2,5	20	10.708

and dust around the NPP where the explosion occurred, while the lighter ones were carried by the wind over Russia, Ukraine, Belarus, Scandinavia, and Europe. Graphs showing the widespread accumulation of Cesium-137 on the surface of Europe as a result of the Chernobyl accident are available on the World Nuclear Association website (World Nuclear Association, 2024). Radiation level indicators at the Forsmark Nuclear Power Plant in Sweden, located 1,000 kilometres away from the Chernobyl plant, sounded alarms on April 28. Authorities at Forsmark reported the situation to the Swedish Radiation Safety Authority and indicated that the radiation originated from outside NPP. On the same day, the Swedish government contacted the Soviet government to inquire whether there had been a nuclear accident in the Soviet Union. Initially, the Soviets denied it, but after the Swedish government claimed, they were preparing to officially alert the International Atomic Energy Agency, the Soviet government acknowledged that an accident had occurred at Chernobyl (Högselius et al., 2024). Thus, only 2 days after the accident, Soviet authorities shared with the world that it was a minor accident and started evacuation procedures around the NPP. On the same day, at 21:02, the news of an accident at the Chernobyl NPP and damage to one unit was broadcast to the entire world through a 20-second television announcement (Nuclear Monitor, 2011).

At the time of the Chernobyl NPP accident, there were 4 units. The first and second units were built between 1970 and 1977, while the third and fourth units had started operating three years before the accident. In the southeast of the facility, an artificial lake of approximately 22 square kilometres was constructed near the Pripyat River, a tributary of the Dnieper River, to provide cooling water to the reactors. One day before the accident, prior to a routine shutdown, the reactor team at Chernobyl Unit 4 began preparing for a test to determine how long the turbines would continue to spin and provide power to the main circulation pumps after the main electrical power supply had been cut off. This test had been conducted the previous year at Chernobyl, but there were concerns about the rapid depletion of power from the turbine, prompting the need to test new voltage regulator designs (World Nuclear Association, 2024). In these tests, which continued on the morning of April 26, it was documented in post-accident investigations that the operators carried out some actions without sufficient knowledge and without paying attention to the safety controls in accordance with the rules. Therefore, when the operator moved to shut down the reactor, the reactor was in an extremely unstable state, and the quirk in the design of the control rods caused a dramatic power increase after insertion into the reactor. As a result of an uncontrollable power surge, the fuel, already very hot along with the cooling water, underwent rapid disintegration, generating steam quickly and causing an increase in pressure. This culminated in the nearly complete destruction of the reactor with two consecutive powerful explosions.

On the day of the accident, there were 600 workers at the NPP. As a result of the explosions, 2 workers died. 134 nuclear power plant personnel and emergency workers were reported to have been exposed to high doses of radiation, resulting in acute radiation syndrome (ARS). Unfortunately, 28 of them later died from ARS (Canadian Nuclear Safety Commision, 2022). Although the incidence of cataracts and leukaemia was high over the years among the employees at the time of the accident, there was no increase in the incidence of leukaemia or solid tumours in general among the employees. It is estimated that the deceased firefighters received doses as high as 20,000 mGy (milligray). Approximately 15,000 persons from the NPP or surrounding communities were reported to have lost their ability to work as a result of diseases which they claimed could be attributed to radiation exposure. Pripyat, which had 45,000 residents, was evacuated on April 27th. By May 14th, approximately 116,000 people living within a 30-kilometre radius had been evacuated and subsequently relocated elsewhere. About 1,000 of them unofficially returned to live in contaminated areas. While most of those evacuated received radiation doses below 50 mSv, a few received doses of 100 mSv or higher. In the subsequent years, 220,000 people were relocated to less contaminated areas, and the initial 30-kilometre radius exclusion zone which is 2,800 km^2 was expanded to cover 4,300 square kilometres.

3.3. Three Mile Island

The construction of the Three Mile Island NPP was completed in 1974 on the Susquehanna River near the state capital of Harrisburg, Pennsylvania. A second reactor at the NPP site, which was very popular because it produced affordable and reliable energy during energy crises, was commissioned in 1978. On the morning of March 28, 1979, at around 4 am, the United States experienced its most significant nuclear power plant accident at this reactor (Barquinero et al., 2021). It was later understood that there had been a release of radioactive gases and radioactive iodine from the reactor into the environment. A series of apparent errors and equipment malfunctions, combined with some suspicious device readings, led to the loss of reactor coolant fluid, overheating of the core, damage to the fuel, but likely no meltdown, and limited release of radioactive noble gases and iodine outside the facility (Corey et al., 1979). The INES rating of this accident has been assessed as 5.

The NPP manager declared an emergency in the plant area, and within half an hour, the station manager announced a general emergency declaration for the NPP at 6:56 am that morning. On that morning, due to conflicting statements about radioactive emissions, the state government closed schools, urged residents to stay indoors, and advised farmers to keep their animals indoors and feed them in sheltered areas. After the accident, much of the effort understandably focused on the amount of radioactive release. It was determined that approximately 2.5 million curies (93 PBq) of radioactive gases and about 15 curies (560 GBq) of iodine-131 were released into the environment. According to the American Nuclear Society, the average radiation dose for people living within 10 miles of the plant was 0.08 mSv, and it did not exceed 1 mSv for any single individual (Nuclear Regulatory Commission, 1979). The accident was a revolution for nuclear power plant construction in the United States and laid the foundation for implementing numerous safety standards.

Conclusion

Over the past 70 years, numerous radiation accidents have occurred due to radiation-producing devices, radioactive materials, and uncontrolled NPP sources. Many people in different parts of the world lost their lives after exposure to internal and external radiation. Many people still continue their lives with various diseases due to radiation exposure. Especially in accidents with high INES values, determining the doses received by workers, emergency response teams, and the public plays an important role in establishing treatment methods (Barquinero J.F., 2021). In addition, the experiences gained from the Chernobyl, and Fukushima nuclear accidents have once again demonstrated the necessity of quickly, and accurately conducting environmental dose measurements to make emergency decisions, initiate evacuation efforts, and implement protective measures promptly. In all of these accidents, it is evident that there have been discussions regarding the method and speed of making these decisions.

There are many rules at different levels in countries to reduce the risk of radiation harm. Although many of these regulations have been prepared based on the recommendations of various experienced international organisations on the subject, there have still been two major accidents with an INES level of 7, posing a high risk of harm to both people and the environment over the past 50 years. It is crucial to carefully and critically evaluate each of these accidents and conduct retrospective studies to determine their causes and implement the specified corrective measures to prevent their recurrence. It is essential to ensure that radiation protection experts and everyone working with radiation are aware of their vital roles. They should be provided with training courses, educational materials, and informational sessions tailored to their professional needs.

References

Alexakhin, R., Anspaugh, L., Balonov, M., Batandjieva, B., Besnus, F., Biesold, H., & Woodhead, D. (2006). Environmental consequences of the Chernobyl accident and their remediation: twenty years of experience. *Report of the Chernobyl Forum Expert Group "Environment."* International Atomic Energy Agency.

Asano, T., Sato, K., & Onodera, J. I. (2001). United Nations scientific committee on the Effects of atomic radiation 2000 report. *Japanese Journal of Health Physics*, *36*(2), 149-158.

Barquinero, J. F., Fattibene, P., Chumak, V., Ohba, T., Della Monaca, S., Nuccetelli, C., & Cardis, E. (2021). Lessons from past radiation accidents: Critical review of methods addressed to individual dose assessment of potentially exposed people and integration with medical assessment. *Environment International*, *146*, 106175.

BBC news. (2011). Timeline: nuclear plant accidents. https://www.bbc. com/news/world-13047267.

Bhargavan, M. (2008). Trends in the utilization of medical procedures that use ionizing radiation. *Health physics*, *95*(5), 612-627.

Canadian Nuclear Safety Commision. (2022). Healty Effect of the Chernobly Accident. https://www.cnsc-ccsn.gc.ca/eng/resources/health/health-effects-chernobyl-accident/.

Carpeggiani, C., Kraft, G., Caramella, D., Semelka, R., & Picano, E. (2012). Radioprotection (un) awareness in cardiologists, and how to improve it. *The international journal of cardiovascular imaging*, *28*, 1369-1374.

Coeytaux, K., Bey, E., Christensen, D., Glassman, E. S., Murdock, B., & Doucet, C. (2015). Reported radiation overexposure accidents worldwide, 1980-2013: a systematic review. *PloS one*, *10*(3), e0118709.

Corey G. R. (1979). A Brief Review of the Accident at Three Mile Island. *IAEA Bulletin*, *21*(5), 54-59.

Donya, M., Radford, M., ElGuindy, A., Firmin, D., & Yacoub, M. H. (2015). Radiation in medicine: Origins, risks and aspirations. *Global Cardiology Science and Practice*, *2014*(4), 57.https://doi.org/10.5339/gcsp.2014.57.

European Society of Radiology (ESR) communications@ myESR. org. (2015). Summary of the European Directive 2013/59/Euratom: essentials for health professionals in radiology. *Insights into imaging*, *6*, 411-417.

Fernández-Arias, P., Vergara, D., & Orosa, J. A. (2020). A global review of PWR nuclear power plants. *Applied Sciences*, *10*(13), 4434.

Ho, M., Obbard, E., Burr, P. A., & Yeoh, G. (2019). A review on the development of nuclear power reactors. *Energy Procedia*, *160*, 459-466.

Hasegawa, A., Ohira, T., Maeda, M., Yasumura, S., & Tanigawa, K. (2016). Emergency responses and health consequences after the Fukushima accident; evacuation and relocation. *Clinical Oncology*, 28(4), 237-244.

Högselius, P., & Klüppelberg, A. (2024). *The Soviet Nuclear Archipelago: A Historical Geography of Atomic-Powered Communism* (p. 178). Central European University Press.

IAEA. (2017). "Uses of Ionizing Radiation for Tangible Cultural Heritage Conservation." *IAEA Radiation Technology Series No. 6*, IAEA, Vienna.

Ichiseki H. (2013). Features of disaster-related deaths after the Great East Japan Earthquake. *The Lancet*, 381(9862), 204.

Independent, N. A. (2011). The Fukushima Nuclear Accident Independent Investigation Commission. *The National Diet of Japan Tokyo, Japan*.

Mettler, F. A. (2012). Medical effects and risks of exposure to ionising radiation. *Journal of Radiological Protection*, *32*(1), N9.

National Academies of Sciences, Engineering, and Medicine. (2021). Radioactive Sources and Alternative Technologies in Industrial Applications. In *Radioactive Sources: Applications and Alternative Technologies*. National Academies Press (US). https://doi.org/10.17226/26121.

Nénot J. C. (2009). Radiation accidents over the last 60 years. *Journal of Radiological Protection*, 29(3), 301.

Nuclear Monitor. (2011). Chernobyl; Chronology of a Disaster. https://www.nirs. org/wp-content/uploads/mononline/nm724.pdf.

Nuclear Regulatory Commission. (1979). Three Mile Island: a Report to the Commissioners and to the Public. Volume II, Part 1 (No. NUREG/CR--1250-(VOL. 2) (PT. 1)). *Nuclear Regulatory Commission*.

Pletcher, K., & Rafferty, J. P. (2016). Japan earthquake and tsunami of 2011. *Encyclopædia Britannica. Encyclopedia Britannica, Inc, 22.* https://www.britannica.com/event/Japan-earthquake-and-tsunami-of-2011.

Raj, B., Jayakumar, T., & Rao, B. P. C. (1995). Non-destructive testing and evaluation for structural integrity. *Sadhana*, *20*, 5-38.

Raskob, W., Dechy, N., Donovan, A., Gallego, E., Nanba, K., Romão, X., ... & Ceballos, H. (2020). Fukushima Daiichi accident in 2011. *Science for Disaster Risk Management 2020-Acting Today, Protecting Tomorrow*. https://doi.org/10.2760/571085.

Reed, A. B. (2011). The history of radiation use in medicine. *Journal of vascular surgery*, *53*(1), 3S-5S. https://doi.org/10.1016/j.jvs.2010. 07.024.

Reisz, J. A., Bansal, N., Qian, J., Zhao, W., & Furdui, C. M. (2014). Effects of ionizing radiation on biological molecules—mechanisms of damage and emerging methods of detection. *Antioxidants & redox signaling*, *21*(2), 260-292.

Saenger E.L. (1986). Radiation accidents. *Annals of Emergency Medicine*, 15(9), 1061-1066.

Sovacool, B. K. (2010). A critical evaluation of nuclear power and renewable electricity in Asia. *Journal of Contemporary Asia*, *40*(3), 369-400.

Sovacool, B. K. (2008). The costs of failure: A preliminary assessment of major energy accidents, 1907–2007. *Energy Policy*, *36*(5), 1802-1820.

Sovakool, B. K. (2009). The accidental century–prominent energy accidents in the last 100 years. *Exploration & Production*, *7*(2), 132-137.

UNCEAR. (2013). Report to the general assembly, scientific annex A: levels and effects of radiation exposure due to the nuclear accident after the 2011 great East-Japan earthquake and tsunami. *Volume I.* https://www.unscear.org/ unscear/uploads/documents/unscear-reports/UNSCEAR_2013_Report_Vol.I. pdf.

UNCEAR. (2022). Report to the general assembly, scientific annex B: Levels and Effects of Radiation Exposure due to the Accident at the Fukushima Daiichi Nuclear Power Station: Implications of Information Published Since the UNSCEAR 2013 Report. *Volume II. https://www.unscear.org/unscear/uploads/documents/unscear-reports/UNSCEAR_2020_21_Report_Vol.II CORR.pdf.*

Wheatley, S., Kröger, W., Chen, L., & Sornette, D. (2018). Open Comprehensive Nuclear Events Database. In *Proc. 14th Intl. Conf. Probabilistic Safety Assessment and Management (PSAM 14).*

World Nuclear Association. (2024). Fukushima Daiichi Accident. https://world - nuclear.org/information-library/safety-and-security/safety-of-Plants/fukushima-daiichi-accident.

World Nuclear Association. (2024). Chernobyl Accident 1986. https://world-nuclear.org/information-library/safety-and-security/safety-of-plants/chernobyl-accident#the-1986-chernobyl-accident.

Chapter 13

Psychological Impact of Radiation Accidents

Cansu Bayğın[1]
and Börte Gürbüz Özgür[2,*]
[1]Psychiatry, Manisa Mental Health and Diseases Hospital, Manisa, Türkiye
[2]Child and Adolescent Psychiatry, Aydin Adnan Menderes University, Aydin, Türkiye

Abstract

Radiation accidents have the potential to affect a wide population, resulting in numerous losses including loss of life, employment, and social connections. Although nuclear disasters are infrequent, their psychological effects are pervasive and long-lasting. According to the 2006 Chernobyl Forum Report, the impact of the Chernobyl disaster on mental health has been recognised as the largest public health problem attributable to a single accident to date. Radiation accidents are linked to various psychological conditions, including depression, anxiety, post-traumatic stress disorder, and medically unexplained somatic symptoms. Behavioural consequences observed in the population include increased suicide rates, sleep disturbances, and heightened use of tobacco and alcohol. Furthermore, stress responses triggered by disasters can exacerbate pre-existing mental health conditions that were stable before the incident. Individuals with greater psychological resilience and social support experience lower rates of psychological distress. Among the most significantly affected by nuclear disasters, both directly and indirectly, are children. In addition to radiation exposure, children face challenges such as parental factors, forced evacuations, family separations, changes in peer and friendship networks, and disruptions to their education. Studies indicate that

---

* Corresponding Author's Email: borte.gurbuz@adu.edu.tr.

In: Triage in Radiation Accidents
Editors: Nural Öztürk and Sibel Karaca
ISBN: 979-8-89530-415-0

children who have experienced a nuclear disaster may continue to suffer from related psychological issues many years later. While there is an extensive body of literature on the physical consequences of radiation exposure, the psychological outcomes have been less thoroughly examined. As a result, in recent years, there has been an increasing interest in research in the field of disaster psychiatry, especially in studies investigating the long-term course of psychiatric symptoms after such events. This section covers the course of psychiatric symptoms that develop after radiation exposure.

Keywords: children, mental health, posttraumatic stress disorders, radiation accidents

1. Introduction

A radiation accident is characterised by exposure to radiation doses exceeding permissible limits or the presence of radioactive contamination during operations involving ionising radiation. The International Atomic Energy Agency (IAEA) defines it as "an event that has led to significant consequences for people, the environment, or the facility." The increasing use of radiation and radioactive materials in contemporary times has led to a rise in such accidents. The most notable example of a major nuclear accident is the 1986 Chernobyl disaster, which involved damage to a reactor core and the release of substantial amounts of radioactivity (Yablokov et al., 2009). The Three Mile Island incident, the Chernobyl nuclear disaster, and the Fukushima nuclear disaster are recognised as the three most significant nuclear accidents in history.

Exposure to radiation resulting from reactor accidents can occur in three primary ways: total or partial body exposure due to proximity to a radiation source, external contamination, and internal contamination. In a radiation accident, all three types of exposure can affect an individual.

2. Acute and Chronic Effects of Radiation

The biological effects of ionising radiation occur when radiation passes through tissue, causing excitation, ionisation, or disruption of molecular structures in the atoms that constitute or are present in the tissue. From a biological perspective, radiation has two types of effects: somatic and genetic. Somatic effects are those that occur in individuals who are directly exposed to radiation. Genetic effects, on the other hand, do not manifest in the exposed individuals but appear in subsequent generations (Daşdağ, 2010).

When considering the duration of exposure, radiation effects can be classified into acute and chronic. Acute effects result from exposure of the entire body or a large part of the body to high doses of radiation over a short period. Chronic effects arise from continuous exposure to low doses of radiation over an extended period (Daşdağ, 2010).

Exposure to high levels of radiation in a short period can lead to acute radiation syndrome (ARS). The primary cause of ARS is the damage to immature parenchymal stem cells in specific tissues. The dose required for ARS development typically ranges between 2 Gray (Gy) and 100 Gy. The first description of ARS was provided by De-Coursey in 1948, following the atomic bomb explosions in Japan during World War II (De Coursey, 1948). ARS involves symptoms affecting three main systems: the gastrointestinal system, the haematopoietic system, and the cerebrovascular system. In addition to these, cutaneous radiation syndrome has also been recently identified. The type of symptoms and the radiation dose causing these symptoms vary among these syndromes.

Acute Radiation Syndrome (ARS) progresses through four distinct phases: the prodromal phase, the latent phase, the manifest illness phase, and either recovery or death.

Prodromal Phase: Symptoms such as nausea, vomiting, diarrhoea, irritability, confusion, and loss of consciousness can occur. This phase develops within minutes to days following radiation exposure, with symptoms lasting from a few minutes to several days.

Latent Phase: This phase is characterised by a period of apparent well-being, which can last from several hours to several weeks.

Manifest Illness Phase: Symptoms specific to the particular syndrome appear during this phase, which can last for several months. Haematopoietic syndrome occurs at doses ranging from 1 to 8 Gy, gastrointestinal syndrome appears at doses between 5 and 20 Gy, and cerebrovascular syndrome develops at doses exceeding 20 Gy.

Recovery or Death: The recovery process can extend from a few weeks to up to two years. Most patients succumb to the effects of radiation exposure within a few months (Centers for Disease Control and Prevention). Localised radiation exposure can cause a range of skin lesions, including hair loss, nail loss, erythema, blistering, desquamation, and necrosis, collectively known as cutaneous radiation syndrome (Waselenko et al., 2004). During the Chernobyl disaster, all 134 patients with acute radiation sickness experienced bone marrow depression, 19 developed widespread radiation dermatitis, and 15 suffered from severe gastrointestinal complications (Mettler et al., 2007).

Long-term effects of radiation exposure include radiation-induced cancer, genetic mutations in subsequent generations, late-onset organ effects (such as typical vascular changes, fibrotic atrophy, and thyroid dysfunction), cataracts, and infertility. The likelihood of developing cancers of the bone marrow, thyroid, breast, and lungs is notably higher with radiation exposure (Türkiye Atom Enerjisi Kurumu, 2009; Uzal & Çaloğlu, 2002). In Hiroshima and Nagasaki, survivors of the atomic bombings showed increased incidences of blood, breast, and other cancers (Ozasa et al., 2012).

3. Trauma

Trauma is defined as an extraordinary event involving actual or threatened death, serious injury, or a threat to physical integrity, experienced by oneself or witnessed happening to a close person (APA DSM-5 Task Force, 2013). Developmental trauma is characterised by its occurrence during critical developmental periods (early childhood or adolescence), its repetitive, prolonged, or cumulative nature, and its perpetration by human hands, particularly through neglect or abuse (physical, sexual, or emotional) by a caregiver during childhood (Courtois, 2004).

Post-Traumatic Stress Disorder (PTSD) is a syndrome that manifests after exposure to a traumatic event, characterised by symptoms of re-experiencing, avoidance, numbing, alienation, and hyperarousal (APA DSM-5 Task Force, 2013). The aetiology of PTSD is believed to involve multiple factors, with a significant focus on the neuroendocrine system, the Hypothalamic-Pituitary-Adrenal (HPA) axis, neurotransmitter systems, and neuroanatomic-neurophysiological models (Mundy & Baum, 2004).

4. Psychological Effects of Radiation Accidents in Adults

Although nuclear disasters have been rare throughout history, their psychological effects are widespread and long-lasting. Victims of such disasters tend to fear man-made catastrophes, especially radiation accidents, more than natural disasters (Neria et al., 2008). Research into the psychological effects of nuclear disasters began following the 1979 Three Mile Island accident in Pennsylvania. Studies on the Chernobyl disaster and the Three Mile Island nuclear power plant accident indicate that the release of radioactive materials instilled significant fear and uncertainty among the public (Bromet, Havenaar, et al., 2011).

Radiation anxiety is characterised by negative thoughts or beliefs concerning the possible harmful health consequences following exposure to radiation (Fukasawa et al., 2017). The primary public health consequence of the Chernobyl disaster is believed to be its profound effect on mental health (Bennett et al., 2006b). The term "radiophobia" was coined after Chernobyl to describe the fear of radiation exposure (Pastel, 2002). Phobias are characterised by an excessive, irrational, and persistent fear of a specific object or situation (APA DSM-5 Task Force, 2013).

In the immediate aftermath of a disaster, many individuals experience insomnia, irritability, and difficulty concentrating (Mellman et al., 1995). Common symptoms also include headaches, dizziness, nausea, fatigue, and weakness (Ford, 1997). These are normal reactions to extraordinary events. Additionally, post-disaster behaviours such as increased use of alcohol, caffeine, and tobacco are known to rise, often representing attempts at self-medication of symptoms (Vlahov et al., 2002).

Disasters create significant trauma for affected individuals due to their magnitude and consequences. Experiencing trauma is considered one of the environmental factors linked to the onset of psychiatric disorders (McQuaid et al., 2001). Traumatic experiences can lead to the development of psychiatric disorders, depending on the frequency, severity, and duration of the traumatic events. Research indicates that 51% of women and 60% of men experience at least one traumatic event in their lifetime (Kessler et al., 1995). Although not all individuals exposed to trauma develop psychiatric disorders, a history of trauma is significant in the aetiology of PTSD, major depression (MD), and substance use disorders (Breslau et al., 1991; Hinton et al., 1997; Jacobsen et al., 2001; Rytwinski et al., 2013).

Individuals exposed to disasters are at increased risk of depression (Miguel-Tobal et al., 2006), generalised anxiety disorder, panic disorder, and heightened substance use (North et al., 2002). Some studies have also shown an increase in suicide rates among disaster-affected populations (Kolves et al., 2013).

Through a detailed examination of post-traumatic stress responses, four distinct groups with different trajectories have been identified:

a) Resilient Group: This group exhibits consistently low levels of PTSD symptoms.
b) Recovery Group: Initially distressed, this group shows gradual remission of symptoms over time.
c) Delayed Reaction Group: Characterised by low symptom levels at the outset, this group experiences a delayed onset of symptoms.
d) Chronic Distress Group: This group maintains consistently high levels of PTSD symptoms over time (Bryant, 2017).

Research indicates that being unmarried, having a lower level of education, and having a lower income make individuals more susceptible to negative mental health outcomes following stressful events (Cronkite & Moos, 1984). Psychological resilience refers to the ability to adapt successfully when faced with stress and adversity. Numerous studies have concluded that psychological resilience serves as a protective factor against the development of psychiatric disorders such as PTSD and depression (Bonanno, 2004).

Nuclear accidents result in PTSD rates ranging from 33.2% to 59.4% within the first-year post-incident (Kukihara et al., 2014; Tsujiuchi et al., 2016). Following the Fukushima nuclear accident, within the first year, PTSD was observed in 14-59% of evacuees (Shiga et al., 2021; Tsujiuchi et al., 2016). A survey conducted in Ibaraki, a neighbouring prefecture of Fukushima, two years after the disaster, revealed that 83.4% of participants exhibited symptoms of depression, and 53.2% showed symptoms of PTSD (Sato et al., 2016). Social isolation and pre-existing physical or mental illness are associated with higher levels of PTSD (Shigemura et al., 2014; Tsujiuchi et al., 2016). Financial concerns and job loss are also linked to PTSD (Tsujiuchi et al., 2016). Individuals forced to evacuate their homes due to nuclear accidents are at higher risk of developing PTSD (Ikeda et al., 2017). A survey conducted one year after the disaster identified chronic physical illnesses, chronic mental disorders, financial concerns, job loss or loss of social connections, and compensation worries as determinants of PTSD (Tsujiuchi et al., 2016). Findings indicate that anxiety and stress-related disorders such as PTSD are more likely in women after disasters, whereas substance-related disorders are more common in men (Bangasser & Valentino, 2014).

Disasters stemming from intentional actions, such as nuclear exposure, typically produce the most severe psychological symptoms (Ursano et al., 2017). Radiation accidents are associated with diseases such as depression, anxiety, PTSD, and medically unexplained somatic symptoms, as well as behavioural consequences such as increased suicide rates, sleep disorders, and increased use of tobacco and alcohol (Bennett et al., 2006a; Bromet, 2012; ICRP Task Group 84, 2012; Kawakami et al., 2020; Terayama et al., 2021; Yabe et al., 2014). Multiple studies have suggested that radiation exposure, as estimated by proximity to the radiation source, is a prevalent risk factor for mental illness (Oe et al., 2021). Exposure to human-made disasters has been associated with an increased risk of lifetime alcohol abuse disorder, PTSD, OCD, and MD. Various studies have reported an increase in PTSD, depression, schizophrenia, and suicidal ideation following radiation disasters (Hori et al., 2023; Shigemura et al., 2021; Sodeyama et al., 2022). Compared to natural disasters, PTSD is more prevalent in human-made disasters (Neria et al., 2008). Oe et al. found that three years after the Fukushima Daiichi nuclear power plant accident, the prevalence of general psychological distress among residents in the evacuation zone was elevated in both men and women compared to the usual levels observed in Japan (Oe et al., 2016). A survey conducted in 2012 with 180,604 individuals evacuated from the Fukushima nuclear disaster found that the perception of radiation risks was associated with psychological distress (Suzuki et al., 2015).

Studies have shown an increase in depression and suicidal thoughts in individuals exposed to high levels of radiation (Loganovsky et al., 2008). Research conducted in Fukushima revealed a prevalence of depression of 14.6% in 2012 (Maeda & Oe, 2017). Individuals with negative perceptions about radiation and its health consequences are more likely to exhibit depressive symptoms (Suzuki et al., 2015). In one study, the prevalence rate of possible depression among adults evacuated from Fukushima was found to be 14.6% in 2012, 11.9% in 2013, and 9.7% in 2014 (Oe et al., 2016). Between 7.1% and 23% of those affected by radiation accidents were reported to meet the criteria for depression diagnosis within the first year ((Lebowitz, 2016; Tsubokura et al., 2014). A study carried out between 1999 and 2002 following the Chernobyl accident showed a depression rate of 13% among 115,191 adolescents, with 77% of the diagnosed individuals being girls (Contis & Foley, 2015). Utilising data from the Pregnancy and Birth Survey of the Fukushima Health Management Survey, it was discovered that 28% of the sample, which comprised 2,262 women, screened positive for depression in 2011 (Goto et al., 2015). In Fukushima, four years after the accident, the frequency of depressive symptoms among female students was reported to be 46.5% (Ito et al., 2018). In this study, higher radiation risk perceptions and reduced efficacy related to reproduction were reported as risk factors contributing to a decline in self-esteem and self-efficacy. Furthermore, proximity to the nuclear power plant and radiation exposure were linked to depression. Women residing closer to the nuclear reactor reported elevated levels of depressive symptoms (Goto et al., 2015).

Tang et al. identified risk factors for depression caused by disasters as being female gender, unmarried status, religious beliefs, low educational level, prior trauma history, experiencing fear during the disaster, and experiencing injury and/or death (Tang et al., 2014). In newly postpartum women, younger maternal age, primiparity, history of past abortion, psychiatric history, obstetric complications, and baby abnormalities were all found to be associated with depressive symptoms (Giannandrea et al., 2013; Mori et al., 2011; Solberg et al., 2011; Tamaki et al., 1997).

Nuclear accidents have been associated with increased suicide rates among decontamination workers after the Chernobyl disaster, but no suicide cases were reported among evacuees (Bromet, 2014). Two years after the Fukushima nuclear accident, it was reported that 8.9% of public employees were at risk of suicide (Maeda et al., 2016). In Fukushima, the overall suicide rate was above the community average in the first year after the disaster, then showed a decline, and three years later showed a rising trend again (Maeda & Oe, 2017; Ohto et al., 2015; Orui et al., 2018). It was found that disaster-related suicide rates were higher in men than in women after the Fukushima disaster (Takebayashi et al., 2020). Risk factors for post-disaster suicidal ideation include psychiatric disorders such as MD and PTSD (Caldera et al., 2001; Chou et al., 2007; Wagenaar et al., 2012), low-income, disaster-related stress (Kessler et al., 2008) and female gender (Stratta et al., 2012; Suzuki et al., 2011).

A cross-sectional study conducted on adults in Japan revealed that 23% of participants expressed concerns regarding their daily radiation exposure (Kusama et al., 2018). Particularly, women and parents with children were found to have a higher likelihood of being anxious about radiation exposure (Rasmussen et al., 2020). Between 13% and 29.8% of pregnant women were reported to be concerned about the effects of radiation on their fetus and baby (Ito et al., 2017). Elderly adults in Fukushima demonstrated greater apprehension about the impact of radiation exposure on future generations, whereas reproductive-age adults expressed more concern about delayed health effects on themselves (Suzuki et al., 2015).

A nuclear disaster, particularly among residents in affected areas, may lead to alcohol use disorder. Numerous studies have reported an association between both human-made and natural disasters and increased alcohol consumption (Fullerton et al., 2013; Garrey et al., 2020; Welch et al., 2014). Specifically, research indicates that individuals experiencing trauma may escalate their alcohol consumption to alleviate symptoms and engage in alcohol use for self-medication purposes (Chilcoat and Menard, 2003; DiMaggio et al., 2009). Experiencing social isolation in the aftermath of disasters is identified as a risk factor for alcohol use disorder (Ishikawa et al., 2012).

Disaster-induced stress responses may lead to acute exacerbations of illnesses that were stable before the disaster occurred (Saeed and Gargano, 2022). Studies have shown exacerbations in bipolar I disorder, with the exacerbation more likely to involve manic transitions than depression.

There is compelling evidence indicating that radiation exposure can serve as an environmental trigger for the onset of schizophrenia (Loganovsky and Loganovskaja, 2000). According to Loganovsky and Loganovskaja's assumption, ionising radiation may serve as an environmental trigger that either precipitates the onset of the illness in those predisposed to schizophrenia or actually induces schizophrenia-like disorders. The neural diathesis-stress factor hypothesis of schizophrenia, which suggests that neurobiological genetic vulnerability can be activated by environmental stressors, is viewed as a model for understanding the effects of exposure to ionising radiation. Reports of increased rates of schizophrenia spectrum disorders following exposure to ionising radiation are found after events such as atomic bombings, nuclear weapon testing, the Chernobyl disaster, environmental contamination with radioactive waste, radiotherapy, and also in areas with a high natural radioactive background. The development of schizophrenia spectrum disorders in individuals excessively exposed to radiation in Chernobyl may be related to the neurophysiological basis of schizophrenia-like symptoms caused by radiation-induced dysfunction of the left frontotemporal limbic function (Loganovsky & Loganovskaja, 2000). Children exposed to prenatal radiation in the Chernobyl accident, especially those exposed during the second trimester of pregnancy, are at higher risk for schizophrenia (Loganovsky & Loganovskaja, 2000; Nyagu et al., 1998).

Traumatic experiences or stressful life events can trigger the onset of obsessive-compulsive disorder (OCD) or increase the risk of its occurrence. Research has indicated a significant increase in the prevalence of OCD following exposure to trauma (O'Donnell et al., 2016). Phenomenologically, the repetitive and persistent thoughts that constitute the obsessions criterion of OCD may resemble the intrusive thoughts in PTSD (Dykshoorn, 2014). A case of a young woman developing OCD due to increased anxiety about radiation contamination following the Fukushima nuclear accident has been reported (Inoue et al., 2017).

Psychological stress can lead to short-term cognitive impairments (Scott et al., 2015). It is generally accepted that high doses of ionising radiation can disrupt cognitive functions in adults particularly in the developing brain. However, many researchers have concluded that there is no relationship between radiation exposure and cognitive impairment (Bromet and Havenaar, 2007). There has been no observed relationship between radiation dose and vascular dementia or Alzheimer's disease in individuals born in Hiroshima before September 1932 (Yamada et al., 2016). Similarly, no differences in overall cognitive functions have been reported in survivors of the Nagasaki atomic bomb (Kinoshita et al., 2019).

Interest in studies investigating the long-term trajectory of psychiatric symptoms in disaster psychiatry is increasing. Norris et al. categorised six trajectories of outcomes following major disasters: resilience, resilience, recovery, recurring/improving, delayed dysfunction, and chronic dysfunction (Norris et al., 2009). Another classification by Bonanno and Diminich identified six trajectories as continuous, chronic, delayed, recovery, improved, and minimally effective resilience (Bonanno & Diminich, 2013).

The impacts of a significant nuclear accident on societies are varied and lasting. Countermeasures should encompass initiatives to enhance disaster management, provide long-term public health services, offer mental and psychological support, provide behavioural and social assistance, and implement measures to mitigate health effects linked to radiation exposure (Ohtsuru et al., 2015).

5. Effects of Radiation Accidents on Child and Adolescent Mental Health

Radiation accidents, affecting many individuals simultaneously from different perspectives, are disasters that can cause major trauma to society, akin to disasters or wars. Disasters affecting society on a large scale can lead to the deterioration of individuals' physical or mental health. They disrupt individuals' access to basic needs such as housing, food supply, and access to clean water. All age groups are directly or indirectly exposed to the adverse effects of this involuntary and rapid change in individuals' lives. Nuclear disasters in history have caused traumatic effects not only in the regions where they occurred but also globally. Such a significant trauma can lead to some psychiatric disorders and neurophysiological changes that affect emotional development. Post-traumatic reactions can vary significantly depending on age, developmental stage, and variables inherent in the nature of the event (such as its origin, severity, and duration) (Lubit et al., 2003). Various studies have reported that developmental factors such as age, gender, proximity to the affected area, personal injury or loss of a family member, degree of life-threatening danger, individual characteristics, family and social support, as well as the source, severity, and duration of the traumatic event, are associated with vulnerability to post-traumatic symptomatology (Hoven et al., 2019).

Nuclear disasters exhibit similarities to natural disasters in terms of their effects. Research on children following natural disasters has consistently shown a range of post-traumatic psychological symptoms, including trauma-specific fears, fears of recurrence, excessive attachment to caregivers, separation anxiety, loss of previously acquired skills, destructive behaviours, post-traumatic repetitive play, behavioural re-enactments, regressive behaviour, feelings of guilt, avoidance of traumatic reminders, school dropout, and somatic complaints. Additionally, other symptoms such as panic attacks, difficulty concentrating, irritability, dysphoria, and substance abuse are frequently reported (Shannon et al., 1994). Among the various presentations of post-traumatic sequelae are PTSD, Acute Stress Disorder, Adjustment Disorders, as well as trauma-related disorders such as attachment issues, Separation Anxiety, Bereavement, Prolonged Grief, Major Depression, and other mood disorders, destructive behaviours, Borderline Personality Disorder, Psychotic Disorders, Somatoform Disorders, and substance abuse (Dunn et al., 1994).

Disasters not only immediately impact children's mental health in response to disaster experiences but also continually impede their cognitive development, which can ultimately affect their long-term development (Kolominsky et al., 1999).

Ionising radiation is considered a potential risk factor for impairment in cognitive functions. Cognitive effects have been observed in individuals exposed to radiation for medical purposes (Greene-Schloesser and Robbins, 2012) and in some individuals affected by the Chernobyl nuclear disaster (Bromet, Havenaar, et al., 2011).

A comprehensive review on this topic has been published recently. This review focuses on the cognitive effects of ionising radiation (IR), both through its direct impact and through the psychological stress induced by IR exposure. High-dose IR exposure has been reported to impair cognitive functions in the developing brain. Although medium and low-dose exposure is potentially harmful to prenatal and neonatal development, there is no consistent evidence indicating that it leads to cognitive impairment in adults (Collett et al., 2020).

In a cohort study involving 1629 children who were exposed to the Chernobyl disaster either in utero or during childhood and subsequently relocated, no overall association was found between the severity of contamination and cognitive function. However, interestingly, children who were in utero at the time of the disaster exhibited significantly higher scores for hyperactivity and attention-deficit/hyperactivity disorder (Bar Joseph et al., 2004).

Another area of investigation regarding the effects of ionising radiation is neurodevelopmental disorders. Since neurodevelopmental disorders begin in the prenatal period and continue into young adulthood, there are suspicions that exposure during this critical period could play a role in their aetiology. Studies on survivors of childhood brain cancer who underwent cranial irradiation have demonstrated neurodevelopmental impairments (Armstrong et al., 2010). However, there is currently no evidence that low and medium-dose exposure leads to neurodevelopmental impairments or causes neurodevelopmental disorders (Pasqual et al., 2020).

Studies investigating the effects of the three major radiation disasters on children's mental health in the short, medium, and long term have been conducted to date. Some studies focusing on the long-term mental effects of the Chernobyl nuclear disaster on children in the region have investigated the risk perceptions of surviving children in the 10 years following the disaster. Overall, these studies found an increase in mood and anxiety symptoms, including post-traumatic stress symptoms, in surviving children compared to controls (Baverstock and Williams, 2006; Bromet, Guey, et al., 2011). Research involving individuals exposed to Chernobyl during infancy has reported significant psychological and behavioural outcomes (Bromet et al., 2005; Bromet et al., 2000; Bromet, Guey, et al., 2011). However, several long-term studies (over ten years) focusing on youth have produced inconsistent results. Bromet and colleagues investigated individuals who were babies or in utero at the time of the Chernobyl disaster and were evacuated from the area 19 years after the disaster for mental health issues such as MD and Generalized Anxiety Disorder (GAD). While there were no significant diagnostic differences compared to the control group, individuals who spent their childhood and adolescence in the shadow of the Chernobyl accident were found to have more intense symptoms related to MD and GAD (Bromet, Guey, et al., 2011).

In more recent times, following the Fukushima nuclear accident, children affected by the disaster or evacuated due to the accident reportedly experienced clinical-level mental health problems in 16% to 25% of cases one year after the Fukushima disaster, according to reports from their parents (Mashiko et al., 2017). The healthy psychological development of infants and children is contingent upon the establishment of positive relationships and interactions with significant adults, such as parents, other family members, and schoolteachers, as well as peers and classmates, with whom they have ongoing interactions. Particularly, the behaviour and attitudes of caregiving parents, who are closest to the child, are among the most crucial factors in psychological development. The parenting behaviours exhibited by a parent who is negatively affected physically and/or mentally by disasters may be adversely impacted. Family fragmentation, forced evacuations, changes in housing and living conditions, and economic changes such as leaving current employment are among the many negative factors that parents must cope with. During the struggle with these challenges, there may be a decrease in the quality of attention, support, and care given to children. This situation can lead to adverse psychological outcomes in both adults and children. This circumstance is a significant risk factor for PTSD in children (Trickey et al., 2012).

Following the Fukushima nuclear disaster, many residents of the affected areas were evacuated voluntarily or involuntarily, and even two years after the disaster, it was observed that 20-40% of evacuated families remained separated from their co-residing family members (Yabe et al., 2014). Additionally, in a study conducted, evacuated mothers reported feeling isolated and struggling with the increased burden of childcare due to the absence of the individuals they usually relied on for childcare (Igarashi et al., 2017).

A study involving mothers with infants indicated that mothers in Fukushima scored higher on measures of psychological distress and postpartum depression compared to mothers in other regions of Japan. It was found that the intense anxiety and depressive symptoms were induced by concerns about radiation exposure during pregnancy and the possibility of the baby being affected by radiation. In a longitudinal study conducted by Kuroda et al., it was found that over 30% of mothers in Fukushima experienced anxiety, and approximately 16-17% exhibited depressive symptoms, including postpartum depression, in the five years following the disaster. These rates were higher than those in other regions of Japan (Kuroda et al., 2021).

Conclusion

Although radiation accidents are relatively rare, when they do occur, they have profound and adverse effects on all living beings in the affected area. These impacts are not confined to the time of the incident but can persist for many years. Exposure to radiation has significant physical consequences and biological effects, as well as substantial psychological implications. Observations and studies conducted on individuals affected by historical radiation accidents underscore the severity of the psychological impact. This psychological impact has been observed across nearly all age groups. In fact, even individuals born after the incident have reported symptoms related to the nuclear disaster. Such major disasters not only have physical effects but also indirectly affect mental health through the social, economic, and cultural consequences they trigger.

The mental health and psychological development of children exposed to a disaster caused by nuclear accidents are affected both directly and indirectly. Although previous studies in this area have provided valuable insights and findings, fully revealing and understanding the actual effects remains challenging. It should be noted that children exposed to a nuclear accident are at risk of anxiety disorders, cognitive dysfunction, PTSD, and depressive symptoms, which require attention and support.

References

APA, DSM-5 Task Force. (2013). *Diagnostic and statistical manual of mental disorders : DSM-5* (5th ed.). American Psychiatric Association.

Armstrong, G. T., Jain, N., Liu, W., Merchant, T. E., Stovall, M., Srivastava, D. K.,... Krull, K. R. (2010). Region-specific radiotherapy and neuropsychological outcomes in adult survivors of childhood CNS malignancies. *Neuro Oncol, 12*(11), 1173-1186. https://doi.org/10.1093/ neuonc/noq104.

Bangasser, D. A., & Valentino, R. J. (2014). Sex differences in stress-related psychiatric disorders: neurobiological perspectives. *Front Neuroendocrinol, 35*(3), 303-319. https://doi.org/10.1016/j.yfrne. 2014.03.008.

Bar Joseph, N., Reisfeld, D., Tirosh, E., Silman, Z., & Rennert, G. (2004). Neurobehavioral and cognitive performances of children exposed to low-dose radiation in the Chernobyl accident: the Israeli Chernobyl Health Effects Study. *Am J Epidemiol, 160*(5), 453-459. https://doi.org/10.1093/ aje/kwh231.

Baverstock, K., & Williams, D. (2006). The chernobyl accident 20 years on: an assessment of the health consequences and the international response. *Environ Health Perspect, 114*(9), 1312-1317. https://doi.org/10.1289 /ehp.9113.

Bennett, B., Repacholi, M., & Carr, Z. (2006a). *Health effects of the Chernobyl accident and special health care programmes Report of the UN Chernobyl Forum Expert Group*. W. H. Organization. https://www.who.int/publications/i/item/9241594179.

Bennett, B., Repacholi, M., & Carr, Z. (2006b). *Health effects of the Chernobyl accident and special health care programmes. Report of the UN Chernobyl Forum Expert Group 'Health'.*

Bonanno, G. A. (2004). Loss, trauma, and human resilience: have we underestimated the human capacity to thrive after extremely aversive events? *Am Psychol, 59*(1), 20-28. https://doi.org/10.1037/0003-066X.59.1.20.

Bonanno, G. A., & Diminich, E. D. (2013). Annual Research Review: Positive adjustment to adversity--trajectories of minimal-impact resilience and emergent resilience. *J Child Psychol Psychiatry, 54*(4), 378-401. https://doi.org/10.1111/jcpp.12021.

Breslau, N., Davis, G. C., Andreski, P., & Peterson, E. (1991). Traumatic events and posttraumatic stress disorder in an urban population of young adults. *Arch Gen Psychiatry, 48*(3), 216-222. https://doi.org/10.1001/ archpsyc.1991.01810270028003.

Bromet, E. J. (2012). Mental health consequences of the Chernobyl disaster. *J Radiol Prot, 32*(1), N71-75.https://doi.org/10.1088/0952-4746/32/1/N71.

Bromet, E. J. (2014). Emotional consequences of nuclear power plant disasters. *Health Phys, 106*(2), 206-210. https://doi.org/10.1097/HP. 0000000000000012.

Bromet, E. J., Gluzman, S. F., Paniotto, V. I., Webb, C. P., Tintle, N. L., Zakhozha, V.,…Schwartz, J. E. (2005). Epidemiology of psychiatric and alcohol disorders in Ukraine: findings from the Ukraine World Mental Health survey. *Soc Psychiatry Psychiatr Epidemiol, 40*(9), 681-690. https://doi.org/10.1007/s00127-005-0927-9.

Bromet, E. J., Goldgaber, D., Carlson, G., Panina, N., Golovakha, E., Gluzman, S. F.,…Schwartz, J. E. (2000). Children's well-being 11 years after the Chornobyl catastrophe. *Arch Gen Psychiatry, 57*(6), 563-571. https://doi.org/10.1001/ archpsyc.57.6.563.

Bromet, E. J., Guey, L. T., Taormina, D. P., Carlson, G. A., Havenaar, J. M., Kotov, R., & Gluzman, S. F. (2011). Growing up in the shadow of Chornobyl: adolescents' risk perceptions and mental health. *Soc Psychiatry Psychiatr Epidemiol, 46*(5), 393-402. https://doi.org/10.1007/ s00127-010-0203-5.

Bromet, E. J., & Havenaar, J. M. (2007). Psychological and perceived health effects of the Chernobyl disaster: a 20-year review. *Health Phys*, *93*(5), 516-521. https://doi.org/10.1097/01.HP.0000279635.14108.02.

Bromet, E. J., Havenaar, J. M., & Guey, L. T. (2011). A 25 year retrospective review of the psychological consequences of the Chernobyl accident. *Clin Oncol (R Coll Radiol)*, *23*(4), 297-305. https://doi.org/10.1016/ j.clon.2011.01.501.

Bryant, R. A. (2017). Acute stress disorder. *Curr Opin Psychol*, *14*, 127-131. https://doi.org/10.1016/j.copsyc.2017.01.005.

Caldera, T., Palma, L., Penayo, U., & Kullgren, G. (2001). Psychological impact of the hurricane Mitch in Nicaragua in a one-year perspective. *Soc Psychiatry Psychiatr Epidemiol*, *36*(3), 108-114. https://doi.org/10.1007/s001270050298.

Centers for Disease Control and Prevention. (23.04.2024). *Acute Radiation Syndrome: Information for Clinicians*. Retrieved 07.07.2024 from https://www.cdc.gov/ radiation-emergencies/hcp/clinical-guidance/ars.html.

Chilcoat, H. D., & Menard, C. (2003). Epidemiological investigations: Comorbidity of posttraumatic stress disorder and substance use disorder. In P. Ouimette & P. J. Brown (Eds.), *Trauma and substance abuse : causes, consequences, and treatment of comorbid disorders* (1st ed., pp. 9–28). American Psychological Association. https://doi.org/https:// doi.org/10.1037/10460-001.

Chou, F. H., Wu, H. C., Chou, P., Su, C. Y., Tsai, K. Y., Chao, S. S.,…Ou-Yang, W. C. (2007). Epidemiologic psychiatric studies on post-disaster impact among Chi-Chi earthquake survivors in Yu-Chi, Taiwan. *Psychiatry Clin Neurosci*, *61*(4), 370-378. https://doi.org/10.1111/j.1440-1819.2007. 01688.x.

Collett, G., Craenen, K., Young, W., Gilhooly, M., & Anderson, R. M. (2020). The psychological consequences of (perceived) ionizing radiation exposure: a review on its role in radiation-induced cognitive dysfunction. *Int J Radiat Biol*, *96*(9), 1104-1118. https://doi.org/10.1080/09553002. 2020.1793017.

Contis, G., & Foley, T. P., Jr. (2015). Depression, suicide ideation, and thyroid tumors among ukrainian adolescents exposed as children to chernobyl radiation. *J Clin Med Res*, *7*(5), 332-338. https://doi.org/10.14740/ jocmr2018w.

Courtois, C. A. (2004). Complex trauma, complex reactions: Assessment and treatment. *Psychotherapy: Theory, Research, Practice, Training*, *41*(4), 412–425. https://doi.org/
https://doi.org/10.1037/0033-3204.41.4.412.

Cronkite, R. C., & Moos, R. H. (1984). The role of predisposing and moderating factors in the stress-illness relationship. *J Health Soc Behav*, *25*(4), 372-393. https://www.ncbi.
nlm.nih.gov/pubmed/6520359.

Daşdağ, S. (2010). Ionizing radiations and cancer. *Dicle Medical Journal*, *37*(2), 177-185.

De Coursey, E. (1948). Human pathologic anatomy of ionizing radiation effects of the atomic bomb explosions. *Mil Surg*, *102*(6), 427-432. https://www.ncbi.nlm.nih. gov/pubmed/18866715.

DiMaggio, C., Galea, S., & Li, G. (2009). Substance use and misuse in the aftermath of terrorism. A Bayesian meta-analysis. *Addiction*, *104*(6), 894-904. https://doi.org/ 10.1111/j.1360-0443.2009.02526.x.

Dunn, G. E., Ryan, J. J., & Dunn, C. E. (1994). Trauma symptoms in substance abusers with and without histories of childhood abuse. *J Psychoactive Drugs*, *26*(4), 357-360. https://doi.org/10.1080/02791072. 1994.10472455.

Dykshoorn, K. L. (2014). Trauma-related obsessive-compulsive disorder: a review. *Health Psychol Behav Med*, *2*(1), 517-528. https://doi.org/10.1080/21642850.2014.905207.

Ford, C. (1997). Somatic symptoms, somatization, and traumatic stress: An overview. *Nordic Journal of Psychiatry*, *51*(1), 5-13. https://doi.org/ 10.3109/08039489709109078.

Fukasawa, M., Kawakami, N., Umeda, M., Miyamoto, K., Akiyama, T., Horikoshi, N.,...Bromet, E. J. (2017). Environmental radiation level, radiation anxiety, and psychological distress of non-evacuee residents in Fukushima five years after the Great East Japan Earthquake: Multilevel analyses. *SSM Popul Health*, *3*, 740-748. https://doi.org/10.1016/ j.ssmph.2017.09.002.

Fullerton, C. S., McKibben, J. B., Reissman, D. B., Scharf, T., Kowalski-Trakofler, K. M., Shultz, J. M., & Ursano, R. J. (2013). Posttraumatic stress disorder, depression, and alcohol and tobacco use in public health workers after the 2004 Florida hurricanes. *Disaster Med Public Health Prep*, *7*(1), 89-95. https://doi.org/10.1017/dmp.2013.6.

Garrey, S. K., Welch, A. E., Jacobson, M. H., Brackbill, R. M., & Gargano, L. M. (2020). The Intentional Self-Medication of 9/11-Related PTSD Symptoms with Alcohol: 15 Years after the Disaster. *Int J Environ Res Public Health*, *17*(15). https://doi.org/10.3390/ijerph17155327.

Giannandrea, S. A., Cerulli, C., Anson, E., & Chaudron, L. H. (2013). Increased risk for postpartum psychiatric disorders among women with past pregnancy loss. *J Womens Health (Larchmt)*, *22*(9), 760-768. https://doi.org/10.1089/jwh.2012.4011.

Goto, A., Bromet, E., & Fujimori, K. (2015). Immediate effects of the Fukushima nuclear power plant disaster on depressive symptoms among mothers with infants: a prefectural-wide cross-sectional study from the Fukushima Health Management Survey. *BMC Psychiatry*, *15*(59). https://doi.org/https://doi.org/10.1186/s12888-015-0443-8.

Greene-Schloesser, D., & Robbins, M. E. (2012). Radiation-induced cognitive impairment--from bench to bedside. *Neuro Oncol*, *14 Suppl 4*(Suppl 4), iv37-44. https://doi.org/10.1093/neuonc/nos196.

Hinton, W. L., Tiet, Q., Tran, C. G., & Chesney, M. (1997). Predictors of depression among refugees from Vietnam: a longitudinal study of new arrivals. *J Nerv Ment Dis*, *185*(1), 39-45. https://doi.org/10.1097/ 00005053-199701000-00007.

Hori, A., Murakami, M., & Tsubokura, M. (2023). Against an insufficient intervention for patients with alcoholism or PTSD: An activity report on a psychiatric clinic after the 2011 complex disaster in Fukushima, Japan. *International Journal of Disaster Risk Reduction*, *96*(103990). https://doi.org/10.1016/j.ijdrr.2023.103990.

Hoven, C. W., Amsel, L. V., & Tyano, S. (2019). *An International Perspective on Disasters and Children's Mental Health* (1st ed. ed.). Springer Nature eBook. https://lib.ugent.be/catalog/ebk01:4100000008737367.

ICRP Task Group 84. (2012). *Initial Lessons Learned from the Nuclear Power Plant Accident in Japan vis-à-vis the ICRP System of Radiological Protection.* https://www.icrp.org/docs/ICRP%20TG84%20Summary%20Report.pdf.

Igarashi, S., Nishizawa, Y., Noto, Y., Kitajima, M., Ogura, N., Sasatake, H., & Hosokawa, Y. (2017). Mothers' Anxieties Concerning Radiation, Child Raising and Quality of Life: A Study of Evacuees of the Fukushima Daiichi Nuclear Disaster. *J. Radiol. Nurs. Soc. Jpn.*, *5*, 3-11.

Ikeda, A., Tanigawa, T., Charvat, H., Wada, H., Shigemura, J., & Kawachi, I. (2017). Longitudinal effects of disaster-related experiences on mental health among Fukushima nuclear plant workers: The Fukushima NEWS Project Study. *Psychol Med*, *47*(11), 1936-1946. https://doi.org/10.1017/ S0033291717000320.

Inoue, K., Inoue, K., & Kato, S. (2017). Obsessive-Compulsive Disorder in the Wake of the Fukushima Nuclear Accident: Anxiety and the Internet. *Prim Care Companion CNS Disord*, *19*(2). https://doi.org/10.4088/ PCC.15l01910.

Ishikawa, K., Kanazawa, Y., Morimoto, S., & Takahashi, T. (2012). Depopulation with rapid aging in Minamisoma City after the Fukushima Daiichi nuclear power plant accident. *J Am Geriatr Soc*, *60*(12), 2357-2358. https://doi.org/10.1111/jgs.12012.

Ito, S., Goto, A., Ishii, K., Ota, M., Yasumura, S., Fujimori, K.,…Birth Survey Group of the Fukushima Health Management, S. (2017). Fukushima Mothers' Concerns and Associated Factors After the Fukushima Nuclear Power Plant Disaster. *Asia Pac J Public Health*, *29*(2_suppl), 151S-160S. https://doi.org/10.1177/1010539516684533.

Ito, S., Sasaki, M., Okabe, S., Konno, N., & Goto, A. (2018). Depressive Symptoms and Associated Factors in Female Students in Fukushima Four Years after the Fukushima Nuclear Power Plant Disaster. *Int J Environ Res Public Health*, *15*(11). https://doi.org/10.3390/ ijerph15112411.

Jacobsen, L. K., Southwick, S. M., & Kosten, T. R. (2001). Substance use disorders in patients with posttraumatic stress disorder: a review of the literature. *Am J Psychiatry*, *158*(8), 1184-1190. https://doi.org/10.1176/ appi.ajp.158.8.1184.

Kawakami, N., Fukasawa, M., Sakata, K., Suzuki, R., Tomita, H., Nemoto, H.,…Bromet, E. J. (2020). Onset and remission of common mental disorders among adults living in temporary housing for three years after the triple disaster in Northeast Japan: comparisons with the general population. *BMC Public Health*, *20*(1), 1271. https://doi.org/10.1186/s12889-020-09378-x.

Kessler, R. C., Galea, S., Gruber, M. J., Sampson, N. A., Ursano, R. J., & Wessely, S. (2008). Trends in mental illness and suicidality after Hurricane Katrina. *Mol Psychiatry*, *13*(4), 374-384. https://doi.org/10.1038/sj.mp.4002119.

Kessler, R. C., Sonnega, A., Bromet, E., Hughes, M., & Nelson, C. B. (1995). Posttraumatic stress disorder in the National Comorbidity Survey. *Arch Gen Psychiatry*, *52*(12), 1048-1060. https://doi.org/10.1001/archpsyc. 1995.03950240066012.

Kinoshita, H., Tanaka, K., Nakao, R., Iso, F., Honda, S., Tanaka, G., & Nakane, H. (2019). Comparison of mental cognitive function of A-bomb survivors and non-A-bomb survivors in Nagasaki. *Psychiatry Clin Neurosci*, *73*(9), 594. https://doi.org/ 10.1111/pcn.12898.

Kolominsky, Y., Igumnov, S., & Drozdovitch, V. (1999). The psychological development of children from Belarus exposed in the prenatal period to radiation from the Chernobyl atomic power plant. *J Child Psychol Psychiatry, 40*(2), 299-305. https://www.ncbi.nlm.nih.gov/pubmed/ 10188713.

Kolves, K., Kolves, K. E., & De Leo, D. (2013). Natural disasters and suicidal behaviours: a systematic literature review. *J Affect Disord, 146*(1), 1-14. https://doi.org/10.1016/
j.jad.2012.07.037.

Kukihara, H., Yamawaki, N., Uchiyama, K., Arai, S., & Horikawa, E. (2014). Trauma, depression, and resilience of earthquake/tsunami/nuclear disaster survivors of Hirono, Fukushima, Japan. *Psychiatry Clin Neurosci, 68*(7), 524-533. https://doi.org/10.1111/
pcn.12159.

Kuroda, Y., Goto, A., Koyama, Y., Hosoya, M., Fujimori, K., Yasumura, S.,...Children's, S. (2021). Antenatal and postnatal association of maternal bonding and mental health in Fukushima after the Great East Japan Earthquake of 2011:The Japan Environment and Children's Study (JECS). *J Affect Disord, 278*, 244-251. https://doi.org/10.1016/ j.jad. 2020.09.021.

Kusama, T., Aida, J., Tsuboya, T., Sugiyama, K., Yamamoto, T., Igarashi, A., & Osaka, K. (2018). The association between socioeconomic status and reactions to radiation exposure: a cross-sectional study after the Fukushima Daiichi nuclear power station accident. *PLoS One, 13*(10), e0205531. https://doi.org/10.1371/journal.pone. 0205531.

Lebowitz, A. J. (2016). Cross-Sectional Data Within 1 Year of the Fukushima Meltdown: Effect-Size of Predictors for Depression. *Community Ment Health J, 52*(1), 94-101. https://doi.org/10.1007/s10597-015-9869-1.

Loganovsky, K., Havenaar, J. M., Tintle, N. L., Guey, L. T., Kotov, R., & Bromet, E. J. (2008). The mental health of clean-up workers 18 years after the Chernobyl accident. *Psychol Med, 38*(4), 481-488. https://doi.org/10.1017/S0033291707002371.

Loganovsky, K. N., & Loganovskaja, T. K. (2000). Schizophrenia spectrum disorders in persons exposed to ionizing radiation as a result of the Chernobyl accident. *Schizophr Bull, 26*(4), 751-773. https://doi.org/10.1093/oxfordjournals.schbul.a033492.

Lubit, R., Rovine, D., DeFrancisci, L., & Eth, S. (2003). Impact of trauma on children. *J Psychiatr Pract, 9*(2), 128-138. https://doi.org/10.1097/ 00131746-200303000-00004.

Maeda, M., & Oe, M. (2017). Mental Health Consequences and Social Issues After the Fukushima Disaster. *Asia Pac J Public Health, 29*(2_suppl), 36S-46S. https://doi.org/
10.1177/1010539516689695.

Maeda, M., Ueda, Y., Nagai, M., Fujii, S., & Oe, M. (2016). Diagnostic interview study of the prevalence of depression among public employees engaged in long-term relief work in Fukushima. *Psychiatry Clin Neurosci, 70*(9), 413-420. https://doi.org/ 10.1111/pcn.12414.

Mashiko, H., Yabe, H., Maeda, M., Itagaki, S., Kunii, Y., Shiga, T.,...Abe, M. (2017). Mental Health Status of Children After the Great East Japan Earthquake and Fukushima Daiichi Nuclear Power Plant Accident. *Asia Pac J Public Health*, *29*(2_suppl), 131S-138S. https://doi.org/10.1177/1010539516675702.

McQuaid, J. R., Pedrelli, P., McCahill, M. E., & Stein, M. B. (2001). Reported trauma, post-traumatic stress disorder and major depression among primary care patients. *Psychol Med*, *31*(7), 1249-1257. https://doi.org/10.1017/s0033291701004202.

Mellman, T. A., Kulick-Bell, R., Ashlock, L. E., & Nolan, B. (1995). Sleep events among veterans with combat-related posttraumatic stress disorder. *Am J Psychiatry*, *152*(1), 110-115. https://doi.org/10.1176/ajp.152.1.110.

Mettler, F. A., Jr., Gus'kova, A. K., & Gusev, I. (2007). Health effects in those with acute radiation sickness from the Chernobyl accident. *Health Phys*, *93*(5), 462-469. https://doi.org/10.1097/01.HP.0000278843.27969.74.

Miguel-Tobal, J. J., Cano-Vindel, A., Gonzalez-Ordi, H., Iruarrizaga, I., Rudenstine, S., Vlahov, D., & Galea, S. (2006). PTSD and depression after the Madrid March 11 train bombings. *J Trauma Stress*, *19*(1), 69-80. https://doi.org/10.1002/jts.20091.

Mori, T., Tsuchiya, K. J., Matsumoto, K., Suzuki, K., Mori, N., Takei, N., & Team, H. B. C. S. (2011). Psychosocial risk factors for postpartum depression and their relation to timing of onset: the Hamamatsu Birth Cohort (HBC) Study. *J Affect Disord*, *135*(1-3), 341-346. https://doi.org/10.1016/j.jad.2011.07.012.

Mundy, E., & Baum, A. (2004). Medical disorders as a cause of psychological trauma and posttraumatic stress disorder. *Current Opinion in Psychiatry*, *17*(2), 123-127. https://doi.org/https://doi.org/10.1097/00001504-200403000-00009.

Neria, Y., Nandi, A., & Galea, S. (2008). Post-traumatic stress disorder following disasters: a systematic review. *Psychol Med*, *38*(4), 467-480. https://doi.org/10.1017/S0033291707001353.

Norris, F. H., Tracy, M., & Galea, S. (2009). Looking for resilience: understanding the longitudinal trajectories of responses to stress. *Soc Sci Med*, *68*(12), 2190-2198. https://doi.org/10.1016/j.socscimed. 2009.03.043.

North, C. S., Tivis, L., McMillen, J. C., Pfefferbaum, B., Spitznagel, E. L., Cox, J.,...Smith, E. M. (2002). Psychiatric disorders in rescue workers after the Oklahoma City bombing. *Am J Psychiatry*, *159*(5), 857-859. https://doi.org/10.1176/appi.ajp. 159.5.857.

Nyagu, A. I., Loganovsky, K. N., & Loganovskaja, T. K. (1998). Psychophysiologic aftereffects of prenatal irradiation. *Int J Psychophysiol*, *30*(3), 303-311. https://doi.org/10.1016/s0167-8760 (98)00022-1.

O'Donnell, M. L., Alkemade, N., Creamer, M. C., McFarlane, A. C., Silove, D., Bryant, R. A., & Forbes, D. (2016). The long-term psychiatric sequelae of severe injury: a 6-year follow-up study. *J Clin Psychiatry*, *77*(4), e473-479. https://doi.org/10.4088/JCP.14m09721.

Oe, M., Fujii, S., Maeda, M., Nagai, M., Harigane, M., Miura, I.,...Abe, M. (2016). Three-year trend survey of psychological distress, post-traumatic stress, and problem drinking among residents in the evacuation zone after the Fukushima Daiichi Nuclear Power Plant accident [The Fukushima Health Management Survey]. *Psychiatry Clin Neurosci*, *70*(6), 245-252. https://doi.org/10.1111/pcn.12387.

Oe, M., Takebayashi, Y., Sato, H., & Maeda, M. (2021). Mental Health Consequences of the Three Mile Island, Chernobyl, and Fukushima Nuclear Disasters: A Scoping Review. *Int J Environ Res Public Health*, *18*(14). https://doi.org/10.3390/ijerph18147478.

Ohto, H., Maeda, M., Yabe, H., Yasumura, S., & Bromet, E. E. (2015). Suicide rates in the aftermath of the 2011 earthquake in Japan. *Lancet*, *385*(9979), 1727. https://doi.org/10.1016/S0140-6736(15)60890-X.

Ohtsuru, A., Tanigawa, K., Kumagai, A., Niwa, O., Takamura, N., Midorikawa, S.,…Clarke, M. (2015). Nuclear disasters and health: lessons learned, challenges, and proposals. *Lancet*, *386*(9992), 489-497. https://doi.org/10.1016/S0140-6736(15)60994-1.

Orui, M., Suzuki, Y., Maeda, M., & Yasumura, S. (2018). Suicide Rates in Evacuation Areas After the Fukushima Daiichi Nuclear Disaster. *Crisis*, *39*(5), 353-363. https://doi.org/10.1027/0227-5910/a000509.

Ozasa, K., Shimizu, Y., Suyama, A., Kasagi, F., Soda, M., Grant, E. J.,…Kodama, K. (2012). Studies of the mortality of atomic bomb survivors, Report 14, 1950-2003: an overview of cancer and noncancer diseases. *Radiat Res*, *177*(3), 229-243. https://doi.org/10.1667/rr2629.1.

Pasqual, E., Bosch de Basea, M., Lopez-Vicente, M., Thierry-Chef, I., & Cardis, E. (2020). Neurodevelopmental effects of low dose ionizing radiation exposure: A systematic review of the epidemiological evidence. *Environ Int*, *136*, 105371. https://doi.org/10.1016/j.envint.2019.105371.

Pastel, R. (2002). Radiophobia: Long-Term Psychological Consequences of Chernobyl. *Military Medicine*, *167*(suppl_1), 134–136. https://doi.org/https://doi.org/10.1093/milmed/167.suppl_1.134.

Rasmussen, J., Ewald, J., & Sterner, T. (2020). Gender and life-stage dependent reactions to the risk of radioactive contamination: A survey experiment in Sweden. *PLoS One*, *15*(4), e0232259. https://doi.org/ 10.1371/journal.pone.0232259.

Rytwinski, N. K., Scur, M. D., Feeny, N. C., & Youngstrom, E. A. (2013). The co-occurrence of major depressive disorder among individuals with posttraumatic stress disorder: a meta-analysis. *J Trauma Stress*, *26*(3), 299-309. https://doi.org/10.1002/jts.21814.

Saeed, S. A., & Gargano, S. P. (2022). Natural disasters and mental health. *Int Rev Psychiatry*, *34*(1), 16-25. https://doi.org/10.1080/09540261. 2022.2037524.

Sato, S., Ishida, I., Hattori, K., & Ota, M. (2016). The psychiatric characteristics of evacuees from Fukushima prefecture in Kitaibaraki city after the Great East Japan Earthquake. *Japanese Journal of Clinical Psychiatry*, *45*, 1457-1464.

Scott, S. B., Graham-Engeland, J. E., Engeland, C. G., Smyth, J. M., Almeida, D. M., Katz, M. J.,…Sliwinski, M. J. (2015). The Effects of Stress on Cognitive Aging, Physiology and Emotion (ESCAPE) Project. *BMC Psychiatry*, *15*, 146. https://doi.org/
10.1186/s12888-015-0497-7.

Shannon, M. P., Lonigan, C. J., Finch, A. J., Jr., & Taylor, C. M. (1994). Children exposed to disaster: I. Epidemiology of post-traumatic symptoms and symptom profiles. *J Am Acad Child Adolesc Psychiatry*, *33*(1), 80-93. https://doi.org/10.1097/00004583-199401000-00012.

Shiga, T., Zhang, W., Ohira, T., Suzuki, Y., Maeda, M., Mashiko, H.,...Fukushima Health Management Survey, G. (2021). Socioeconomic status, damage-related conditions, and PTSD following the Fukushima-daiichi nuclear power plant accident:The Fukushima Health Management Survey. *Fukushima J Med Sci*, *67*(2), 71-82. https://doi.org/10.5387/fms.2020-24.

Shigemura, J., Tanigawa, T., Nishi, D., Matsuoka, Y., Nomura, S., & Yoshino, A. (2014). Associations between disaster exposures, peritraumatic distress, and posttraumatic stress responses in Fukushima nuclear plant workers following the 2011 nuclear accident: the Fukushima NEWS Project study. *PLoS One*, *9*(2), e87516. https://doi.org/10.1371 /journal.pone.0087516.

Shigemura, J., Terayama, T., Kurosawa, M., Kobayashi, Y., Toda, H., Nagamine, M., & Yoshino, A. (2021). Mental health consequences for survivors of the 2011 Fukushima nuclear disaster: a systematic review. Part 1: psychological consequences. *CNS Spectr*, *26*(1), 14-29. https://doi.org/10.1017/S1092852920000164.

Sodeyama, N., Tachikawa, H., Takahashi, S., Aiba, M., Haraguchi, Y., & Arai, T. (2022). The Mental Health of Long-Term Evacuees outside Fukushima Prefecture after the Great East Japan Earthquake. *Tohoku J Exp Med*, *257*(3), 261-271. https://doi.org/10.1620/tjem.2022.J038.

Solberg, O., Dale, M. T., Holmstrom, H., Eskedal, L. T., Landolt, M. A., & Vollrath, M. E. (2011). Long-term symptoms of depression and anxiety in mothers of infants with congenital heart defects. *J Pediatr Psychol*, *36*(2), 179-187. https://doi.org/10.1093/jpepsy/jsq054.

Stratta, P., Capanna, C., Riccardi, I., Carmassi, C., Piccinni, A., Dell'Osso, L., & Rossi, A. (2012). Suicidal intention and negative spiritual coping one year after the earthquake of L'Aquila (Italy). *J Affect Disord*, *136*(3), 1227-1231. https://doi.org/10.1016/
j.jad.2011.10.006.

Suzuki, Y., Tsutsumi, A., Fukasawa, M., Honma, H., Someya, T., & Kim, Y. (2011). Prevalence of mental disorders and suicidal thoughts among community-dwelling elderly adults 3 years after the niigata-chuetsu earthquake. *J Epidemiol*, *21*(2), 144-150. https://doi.org/10.2188/ jea.je20100093.

Suzuki, Y., Yabe, H., Yasumura, S., Ohira, T., Niwa, S., Ohtsuru, A.,...Mental Health Group of the Fukushima health management, s. (2015). Psychological distress and the perception of radiation risks: the Fukushima health management survey. *Bull World Health Organ*, *93*(9), 598-605. https://doi.org/10.2471/BLT.14.146498.

Takebayashi, Y., Hoshino, H., Kunii, Y., Niwa, S. I., & Maeda, M. (2020). Characteristics of Disaster-Related Suicide in Fukushima Prefecture After the Nuclear Accident. *Crisis*, *41*(6), 475-482. https://doi.org/10.1027/0227-5910/a000679.

Tamaki, R., Murata, M., & Okano, T. (1997). Risk factors for postpartum depression in Japan. *Psychiatry Clin Neurosci*, *51*(3), 93-98. https://doi.org/10.1111/j.1440-1819.1997.tb02368.x.

Tang, B., Liu, X., Liu, Y., Xue, C., & Zhang, L. (2014). A meta-analysis of risk factors for depression in adults and children after natural disasters. *BMC Public Health*, *14*, 623. https://doi.org/10.1186/1471-2458-14-623.

Terayama, T., Shigemura, J., Kobayashi, Y., Kurosawa, M., Nagamine, M., Toda, H., & Yoshino, A. (2021). Mental health consequences for survivors of the 2011 Fukushima nuclear disaster: a systematic review. Part 2: emotional and behavioral consequences. *CNS Spectr*, *26*(1), 30-42. https://doi.org/10.1017/S1092852920000115.

Trickey, D., Siddaway, A. P., Meiser-Stedman, R., Serpell, L., & Field, A. P. (2012). A meta-analysis of risk factors for post-traumatic stress disorder in children and adolescents. *Clin Psychol Rev*, *32*(2), 122-138. https://doi.org/10.1016/j.cpr.2011.12.001.

Tsubokura, M., Hara, K., Matsumura, T., Sugimoto, A., Nomura, S., Hinata, M.,…Kami, M. (2014). The immediate physical and mental health crisis in residents proximal to the evacuation zone after Japan's nuclear disaster: an observational pilot study. *Disaster Med Public Health Prep*, *8*(1), 30-36. https://doi.org/10.1017/dmp.2014.5.

Tsujiuchi, T., Yamaguchi, M., Masuda, K., Tsuchida, M., Inomata, T., Kumano, H.,…Mollica, R. F. (2016). High Prevalence of Post-Traumatic Stress Symptoms in Relation to Social Factors in Affected Population One Year after the Fukushima Nuclear Disaster. *PLoS One*, *11*(3), e0151807. https://doi.org/10.1371/journal.pone.0151807.

Türkiye Atom Enerjisi Kurumu. (2009). *Radyasyon, İnsan ve Çevre: İyonlaştırıcı Radyasyon, Etkileri ve Kullanım Alanları, Güvenli Kullanımı İçin Uygulamada Olan Tedbirler*. TAEK Yayınları.

Ursano, R. J., Fullerton, C. S., Weisæth, L., & Raphael, B. (2017). *Textbook of disaster psychiatry* (Second edition. ed.). Cambridge University Press.

Uzal, C., & Çaloğlu, M. (2002). The Role of Ionizing Radiation in the Etiology of Cancer. *Balkan Medical Journal*, *19*(3-4), 177-182.

Vlahov, D., Galea, S., Resnick, H., Ahern, J., Boscarino, J. A., Bucuvalas, M.,…Kilpatrick, D. (2002). Increased use of cigarettes, alcohol, and marijuana among Manhattan, New York, residents after the September 11th terrorist attacks. *Am J Epidemiol*, *155*(11), 988-996. https://doi.org/10.1093/aje/155.11.988.

Wagenaar, B. H., Hagaman, A. K., Kaiser, B. N., McLean, K. E., & Kohrt, B. A. (2012). Depression, suicidal ideation, and associated factors: a cross-sectional study in rural Haiti. *BMC Psychiatry*, *12*, 149. https://doi.org/10.1186/1471-244X-12-149.

Waselenko, J. K., MacVittie, T. J., Blakely, W. F., Pesik, N., Wiley, A. L., Dickerson, W. E.,…Strategic National Stockpile Radiation Working, G. (2004). Medical management of the acute radiation syndrome: recommendations of the Strategic National Stockpile Radiation Working Group. *Ann Intern Med*, *140*(12), 1037-1051. https://doi.org/10.7326/0003-4819-140-12-200406150-00015.

Welch, A. E., Caramanica, K., Maslow, C. B., Cone, J. E., Farfel, M. R., Keyes, K. M.,…Hasin, D. S. (2014). Frequent binge drinking five to six years after exposure to 9/11: findings from the World Trade Center Health Registry. *Drug Alcohol Depend*, *140*, 1-7. https://doi.org/10.1016/ j.drugalcdep.2014.04.013.

Yabe, H., Suzuki, Y., Mashiko, H., Nakayama, Y., Hisata, M., Niwa, S.,…Mental Health Group of the Fukushima Health Management, S. (2014). Psychological distress after the Great East Japan Earthquake and Fukushima Daiichi Nuclear Power Plant accident: results of a mental health and lifestyle survey through the Fukushima Health Management Survey in FY2011 and FY2012. *Fukushima J Med Sci*, *60*(1), 57-67. https://doi.org/10.5387/fms.2014-1.

Yablokov, A. V., Nesterenko, V. B., & Nesterenko, A. V. (2009). *Chernobyl: Consequences of the Catastrophe for People and the Environment* (Vol. 1181). Blackwell Publishing on behalf of the New York Academy of Sciences.

Yamada, M., Landes, R. D., Mimori, Y., Nagano, Y., & Sasaki, H. (2016). Radiation Effects on Cognitive Function Among Atomic Bomb Survivors Exposed at or After Adolescence. *Am J Med*, *129*(6), 586-591. https://doi.org/10.1016/j.amjmed.2015.09.002.

Chapter 14

Emergency Plans and Training in Radiation Accidents

Doğan Yaşar*

Faculty of Arts and Sciences, Physics Department, Kırşehir Ahi Evran University, Kırşehir, Türkiye

Abstract

Nuclear and radiological techniques are widely utilised in energy, health, industry and defense sectors. They pose risks to public health and the environment, while providing significant benefits to human life. The International Atomic Energy Agency (IAEA) has published technical reports aimed at preventing or mitigating the harmful effects of radioactivity and radiation released in a nuclear or radiological accident. In this context, to effectively respond to a nuclear or radiological emergency, sufficient capacity must be developed within the business organisation and at local, regional, national, and when necessary, international levels. The emergency plan outlined in this study has been designed with a practical perspective for national, regional, local and facility/activity levels concerning nuclear or radiological emergency response and prevention. This emergency plan encompasses procedures, equipment and multi-disciplinary professional groups, where authorities and responsibilities are clearly defined. It operates within an organisational framework that ensure coordination and includes training and exercises.

---

* Corresponding Author's Email: dogan.yasar@ahievran.edu.tr

In: Triage in Radiation Accidents
Editors: Nural Öztürk and Sibel Karaca
ISBN: 979-8-89530-415-0

Keywords: contamination, emergency plan, preparedness, radioactivity, radiological response

1. Introduction

Any substance, material or device that can cause radiation exposure is referred to as a radiation source. People are affected by exposure to radioactive substances in two primary ways: external and internal. Certain radioactive nuclides pose the risk of external radiation by emitting their energy in the form of gamma radiation. The risk of radiation exposure increases with time spent near such a source. Additionally, radioactive materials in the form of smoke, dust or liquid contribute to pollution, and the contamination of such substances on a surface, object or person is termed external contamination. Contamination on the skin can lead to severe skin burns. This can occur through contact with radioactive materials that have leaked or spilled from a container. However, unknowingly swallowing food by eating and drinking with contaminated hands may also pose a risk of internal contamination. Radioactive material can also be dangerous if it enters a person's body through inhalation, ingestion, or open wounds. This is called internal contamination (IAEA, 2020a). Therefore, it is imperative to prevent the use and handling of radioactive materials by non-experts and to raise awareness. Inhalation by those in the immediate vicinity as a result of the evaporation of radioactive material due to a fire or explosion that may occur in the area where a radioactive source such as ^{131}I is located, may cause potentially serious health effects (IAEA, 2006a).

Radiation accidents in past years have shown that holding or moving a radiation source for only a few minutes causes serious deterministic health effects (IAEA, 2011; IAEA, 2012; IAEA, 2020a; Pedraza, 2013). A radiological accident can cause serious deterministic health effects. Health effects resulting from changes in cellular functions when living things are exposed to high radiation doses are called "deterministic effects." Examples of deterministic effects include radiation-induced cataracts, hypothyroidism, infertility, and effects on the embryo and foetus (IAEA, 2007; IAEA, 2020a).

Today, nuclear and radiological techniques are increasingly used for military purposes, electricity generation, industry, agriculture and medicine. While using these techniques, they also bring risks to the environment and the public. Countries that use technology containing nuclear materials and radiation must be aware of the risks as well as the benefits they provide and have the responsibility of minimising the negative consequences. It has been experienced that many nuclear or radiological accidents from the past to the present have caused great damage to the environment and the public locally, regionally and even across borders (IAEA, 2012; IAEA, 2006b). Countries that decide to use nuclear and radiological technology are obliged to carry out emergency studies and preparations to minimise damage in the event of an accident. The release of energy or exposure to emitted radiation resulting from a nuclear chain reaction and the decay of reaction products are hazards requiring a nuclear or radiological emergency (IAEA, 2015).

The International Atomic Energy Agency (IAEA) has published technical reports to enable member states to develop, justify and optimise protection strategies in preparation for the effective taking of protective and other response measures in the event of a nuclear or radiological emergency (IAEA, 2014; IAEA, 2015). In these reports, hazards assessed in terms of IAEA security requirements are grouped by major emergency preparedness categories. These are

I. Facilities such as nuclear power plants that are assumed to cause deterministic effects and require urgent intervention in accordance with international standards.
II. Nuclear ship and submarine
III. Industrial irradiation facilities or some hospitals where emergency intervention and protective actions must be carried out on-site in accordance with international standards.
IV. Transportation of mobile nuclear or radioactive materials such as industrial radiography sources, nuclear-powered satellites, or radioisotope generators, theft of a hazardous source, and use of radiological irradiation devices.

Approaches to improving the ability to respond to a nuclear or radiological emergency vary depending on the characteristics of the emergency. It is appropriate to divide emergency preparedness and response into two classes:

a) Nuclear Emergencies

These are classified as threat category I, II or III, depending on on-site and off-site threats. Nuclear emergencies may occur in the following situations (IAEA, 2007);

- Large irradiation facilities
- Nuclear reactors (research reactors, ship reactors and power reactors);
- Facilities for storing large quantities of spent fuel or liquid or gaseous radioactive waste;
- Fuel conversion facilities
- Industrial facilities
- Research or medical facilities with large, fixed resources

b) Radiological Emergencies

These are classified in threat category IV (IAEA, 2007). Radiological emergencies include:

- Uncontrolled unclaimed resources
- Misuse of industrial and medical hazardous resources
- Public exposure and contamination from unknown sources
- Re-entry of a satellite containing radioactive material
- Serious overexposure
- Malicious threats and/or actions
- Transportation emergencies.

These are situations that require emergency preparedness for the uncontrolled release of ionising radiation to the environment or to people from a sealed or unsealed high hazard radioactive source that is widely used in almost all countries. Sources with a possible accident risk can be listed as sealed sources such as 60Co, 137Cs or 192Ir used in medicine and industry, and open radiopharmaceutical sources used for diagnostic and therapeutic purposes in nuclear medicine and scientific research (IAEA, 2020a). In rare cases, X-ray equipment and linear particle accelerators also cause uncontrolled radiation exposure to the public. The radiological emergency plan developed in this study applies to applications requiring emergency response and prevention on site and in facilities for radiological activities specified in Hazard Categories III and IV.

2. Goals of Emergency Response

It is essential to develop a comprehensive emergency plan that takes into account the various negative consequences of a nuclear or radiological accident. In addition to protecting the public from the harmful effects of radiation exposure, it should also be ensured that the negative economic, social and psychological consequences are minimised. In this context, during a nuclear or radiological emergency, it is important to prioritise the following parameters in order to achieve the objectives of the emergency response (IAEA, 2015; IAEA, 2020b).

- Save lives
- Prevent or minimise serious deterministic effects
- Provide first aid and critical medical treatment
- Reduce the risk of stochastic effects
- Inform the public and maintain confidence
- Reduce non-radiological consequences
- Protect property and the environment
- Prepare for the resumption of social and economic activities

3. Concept of Protection Strategy

The protection strategy outlines the national approach to nuclear and radiological emergency response in order to achieve the emergency response objectives. This approach involves the evaluation of preventive and other response measures, individually or together, on the basis of reference levels and general criteria expressed in terms of residual and predicted or received doses (IAEA, 2020b). The protection strategy describes in detail what is to be achieved in response to a nuclear or radiological emergency, from the declaration of the emergency to the end of the emergency. The protection strategy thus forms the basis for the revision of existing emergency regulations and the development of regulations among the responding authorities.

3.1. Scope of a Radiation Emergency Plan

The aim of emergency preparedness is to ensure that the facility and the enterprise are ready to respond effectively to radiological emergencies at the local, regional, and national levels, and even internationally. It also ensures that the national emergency program complies with international emergency regulations (IAEA, 2015).

However, all radiation facilities and activities operating in the country must develop a radiation emergency plan to respond to emergencies involving radiation sources. The scope of emergency preparation and response for each event, the clarity of the plan, and the emergency response organisation should be described in detail.

4. Informing the Public of the Radiation Emergency Plan

The key to successful emergency response is to establish effective communication with the public (IAEA, 2015). The primary objectives of public communication in a nuclear or radiological emergency should include

- Protect public health
- Informing the public about protection and response and the nature of the threat during the preparedness and response phases
- To gain and maintain public confidence in the emergency response through transparent, timely, open and accurate public communication.
- Addressing public concerns about potential adverse impacts on public health, property and the environment
- Preventing unnecessary anxiety, reducing psychological impact
- Tackle misinformation and rumours

5. Emergency Response Organisation

A general organisation should be established for preparation and response to radiological emergencies, and adequate personnel should be employed and assigned suitable tasks. The Emergency Response Organisation (ERO) must focus on preventing_radiation accidents and reduce the consequences of accidents that occur. The roles and responsibilities of each individual within the ERO should be clearly defined in the radiation emergency plan. The assessment of radiological conditions should include arrangements for controlling an uncontrolled radioactive source to avert a hazardous situation and eliminate its negative impacts, and for managing radiation exposure of employees.

It is essential to act swiftly to prevent and reduce the danger in accident situations such as theft or loss of the radiation source, or exposure of the public and employees to high doses (PNRA, 2016). The roles and responsibilities of the emergency manager and each group leader should be clearly defined in the radiation emergency plan.

Activity groups in the formation of the Emergency Response Organisation:

- Emergency coordination
- Facility management and services
- Radiation protection
- Medical services
- Security and access control
- Fire safety and hazardous material control
- Public information

Additionally, the emergency response organisational chart should be included in the plan.

6. Personal Responsibilities of the Emergency Response Organisation

6.1. Emergency Manager

The emergency manager is primarily responsible for planning a timely, organised and effective emergency response. The manager must possess the authority to organise and control emergency response personnel and work closely with off-site response organisations. They also bear the responsibility and authority to initiate any necessary intervention action. In emergency situations, the manager and each group leader must coordinate operations. The head of each functional area should report to the emergency manager.

They ensure effective coordination between the emergency manager, subordinate control groups and response personnel in on-site and off-site interventions and activities (PNRA, 2016).

6.2. Facility Administration and Services

Facility management and services play an important place in the emergency preparedness and response organization. They are tasked with ensuring the continuity of basic services such as electricity, water, ventilation and air conditioning within the facility. These services must be provided without interruption. The group manager should be appointed as a member of the emergency response team. This service group is responsible for related services such as damage assessment, malfunction repair, and technical support. It is important to conduct training and exercises that take into account the risks of radiation and contamination in emergency situations (PNRA, 2016).

6.3. Radiation Control and Safety

The radiation control and safety group are tasked with conducting environmental radiation surveys, maintaining controlled areas, identifying and measuring radiation and contamination areas, and advising personnel on radiation control procedures (NYC, 2009; IAEA, 2018). Radiation control and safety personnel must have knowledge of radiation and radioactivity to assist in the preparation, planning and intervention of facilities. They may be selected from the following professions, depending on the staffing status of the facility (PNRA, 2016; NYC, 2009; IAEA, 2020c).

- Radiation Safety Officer (RSO)
- Medical physicist
- Health physicists
- Nuclear medicine physicians and technologists
- Radiation oncology physicians and technologists
- Radiology Specialists

6.3.1. Radiation Safety Officer

The roles and responsibilities of the radiation safety officer (RSO) including emergency workers, such as medical physicists, health physicists, and radiation protection officers, should be clearly defined. The RSO establishes the appropriate level of protection during the emergency response in accordance with the country reference level. They are responsible for calculating potential doses from radiation exposure and determining suitable working hours for personnel. The RSO must also organise dosimetry, decontamination, and environmental monitoring efforts (IAEA, 2018).

6.3.2. Radiation Control and Security Group, Scope of Duty

- Identifying high radiation sources and isolate, shield or move them as appropriate in accordance with radiation safety procedures.
- Coordinating with facility security regarding access to radiation control areas.
- Shield and secure high activity sources.
- Minimising contamination in control areas and informing personnel working in these areas of dose rates.
- Carrying out work to prevent the spread of contamination outside control areas.
- Assessing patients and staff as frequently as possible for external contamination. (IAEA, 2020c).

6.3.3. Responsibilities of a Radiation Safety Officer

Planning

- The radiation safety officer is responsible for radiological safety, which includes assessing the suitability of activities involving radiation, obtaining activity permits from the national authority, and determining compliance with national quality safety rules and regulations.
- The radiation safety officer or designated person shall conduct training programmes for personnel at risk of receiving high radiation doses in the event of an accident, and shall train them in the use of radiation measurement and monitoring equipment.
- The radiation safety officer or designated person shall perform compliance checks on the calibration of measurement systems and equipment and review the radiation safety programme annually.

Response

- The radiation safety officer or designated person shall supervise the radiation control team and advise the emergency manager on matters relating to the radiological emergency.

Recovery

- The radiation safety officer is responsible for training personnel working in the facility in the safe use, storage and disposal of open and sealed radioactive sources and equipment.
- The radiation safety officer records the doses of all personnel exposed to radiation and informs them of their exposure (IAEA, 2020c).

6.4. Medical/Health Professionals

Medical and health professionals are indispensable to emergency prevention and response organisations. If medical care is unavailable at a facility, suitable hospitals with the capacity to treat radiation-exposed or contaminated individuals should be contacted. The licensee must ensure that the medical and healthcare professional responsible for delivering emergency medical care is provided with the necessary training in emergency response and participates in periodic emergency drills held at the facility (NRRC, 2023).

6.5. Security and Access Control

Facility security personnel must develop a control mechanism that considers the risks and threats that may be encountered in emergency situations, and appropriate training and equipment must be provided. Basic responsibilities of security personnel during emergency response (PNRA, 2016);

- Evacuating facility personnel
- Establishing and controlling assembly areas
- Conducting a head count
- Controlling access

- Arranging transport and communication vehicles for emergency teams
- Liaising with security agencies and search and rescue services during emergency response.
- Restricting access to the affected facility or activity area to employees other than emergency response personnel
- Cordoning the area and hanging warning signs and symbols.

6.6. Fire Safety and Hazard Control Group

The fire safety group manager must have knowledge about the fire potential of the facility, its layout, fire protection systems and equipment, personnel training and qualifications. Training for the fire protection group should be conducted through emergency drills at regular intervals. Facilities that do not have a full-time fire safety department will depend on the response of the local fire protection department. In this case, the contact information and role of off-site support agencies, such as fire and police, should be included in the facility's emergency plan (NRRC, 2023).

6.7. Public Information Officer (PIO)

The emergency plan should outline information regarding the responsibilities of the public information officer. The public information officer (PIO) is the person responsible for the timely, clear disclosure of essential information about the radiological emergency to the media and the public. This individual_may be a designated official or emergency manager. The emergency plan should include detailed information regarding the responsibilities of the public information officer (NRRC, 2023).

Information provided to the media must be prepared and reviewed by the PIO and published after management approval. The information presented must be simple, understandable, up-to-date and consistent. Must have basic knowledge about radiation and its terminology, which the PIO includes in its press releases and statements (IAEA, 2020d).

7. Managing Medical Response in Radiological Emergency

In case of any radiological accident, health authorities and hospital management should be informed about the possibility of radiation exposure and contamination risk. If an individual is exposed to radiation and clinical signs or symptoms of a possible radiological emergency occur, the competent authority is notified, and the necessary preparations are made.

X-ray machines and linear accelerators cannot cause radioactive contamination. Improper use of gamma sources, or unshielded sources, causes external exposures with significant clinical consequences.

7.1. Initial Treatment

In facilities where a nuclear or radiological accident may occur, the emergency team on site must be trained regularly to provide first medical intervention.

7.2. Treatment in the Reception Area

Medical team members caring for those injured in a radiological accident must have the skills to perform a preliminary assessment, triage, and provide any necessary treatment to persons exposed to radiation. This team should include doctors, nurses, healthcare workers and a physicist with expertise in monitoring and assessing radiation and its contamination (IAEA, 1998).

7.2.1. Preparation of the Reception Area

Special rules and precautions must be taken to prevent the spread of radioactive contamination in patients hospitalised as a result of a radiological accident.

This application will ensure that clean areas are not contaminated and therefore service is provided in treatment areas without interruption.

Before referred patients arrive at the reception area, the following preparations should be made:

- An area should be reserved outside the building where vehicles transporting patients can be checked for contamination and decontaminated when necessary.
- To minimise the potential spread of contamination, a limited number of trained personnel should care for patients.
- The hands, feet, clothing and equipment of the personnel leaving the contaminated area should be monitored without interruption with measuring devices. In case of contamination, a decontamination procedure should be applied.

Protective clothing (waterproof apron, hat, mask, gloves and shoe covers) should be provided to personnel caring for patients.

- Large waterproof nylon bags should be kept for contaminated clothing and disposable materials.
- Paper towels and tissues should be accessible for decontamination.
- Radiation warning signs should be posted to prevent unauthorised entry.
- Special containers should be available for tissue, urine, and stool samples (UNLV, 2018).

7.3. Dosimetry Services

It is essential to assess the level and extent of radiological pollution that threatens responding workers, the public, and the environment if urgent action is not taken in a radiological accident. Radiological pollution requires immediate action due to the danger of both external and internal radiation. Determining the absorbed dose based on the type of ionising radiation that poses a radiation risk in the event of a radioactive fallout, spillage or scattering in an accident is an extremely complex issue. The type of ionising radiation that causes the radiation dose, that is, the level and extent of alpha, beta, neutron or gamma rays, should be determined with special survey instruments. Unless the person exposed to radiation is wearing a personal dosimeter, assessing radiation exposure requires expertise and the boiler will need to be reconstructed to estimate the absorbed dose. For this, a detailed interview with the relevant people is required. Dose assessment can be made using techniques such as electron spin resonance analysis using samples taken from the clothes of people exposed to radiation, any belongings they carry or other materials found in the environment (IAEA, 1998).

7.3.1. Radioactive Contamination Detection Devices

Radiation survey measurements are crucial in an incident involving radionuclide contamination. These measurements are conducted_by a radiation protection specialist or health physicist experienced in instrumentation. Personnel who do not frequently use radiation measurement devices must undergo specialised training. Depending on the type of radiation emitted by the radionuclide causing the contamination, appropriate instruments are used for measurement. Special radiation monitoring equipment is required to measure external contamination and then to study decontamination.

These devices:

- Geiger-Mueller (GM) meters are devices used to detect radiation contamination in people and the environment.
- The pancake probe (Pancake GM) detects alpha, beta or gamma radiation. It is most effective at detecting beta radiation. It is best used to detect low levels of radioactive contamination on people or surfaces.
- A thin-window GM counter or alpha scintillation probe is required to measure alpha particles.
- Electronic (digital) dosimeters can assess an individual's exposure to radiation, also known as personal dosimetry. They can also be used to detect and measure radiation.
- Personal passive detectors are used to measure an individual's total radiation dose over the period of use (TLD, OSL, etc.).
- Dose rate meters with Geiger-Mueller (GM) detectors are used to detect radiation contamination in humans and the environment.
- The pancake probe (pancake GM) can detect alpha, beta or gamma radiation. It is most effective at detecting beta radiation.
- Detectors that can measure alpha particles must have thin windows to allow charged alpha particles to enter the detector. A thin window GM counter or alpha scintillation probe is required for this measurement.
- Digital electronic dosimeters are used as personal dosimeters with dose and dose rate measurement modes.

- Personal passive detectors are used to measure individual total radiation dose over the period of use (TLD, OSL, etc.) (NYC, 2009).

8. Termination of Nuclear or Radiological Emergency

In this section of the radiation emergency plan, the criteria for deciding to terminate the emergency should be determined and explained. The process for ending the emergency should be clearly defined, including communications and announcements to be made outside working and working hours.

9. Analysing Nuclear or Radiological Emergency and Response

It is essential to perform a timely and detailed analysis of the radiological emergency and emergency response with the participation of relevant parties. This study aims to prevent accidents and possible disruptions that may occur during the intervention and will also contribute to internationally coordinated analyses. Due attention should be paid to the following issues in the analysis:

- Reconstruction of emergency conditions
- Cause of the emergency
- Regulations and regulatory controls
- General safety implications
- Emergency improvements

10. Record Keeping and Report Preparation

The emergency plan must have arrangements for keeping and preserving the necessary records and reports. The emergency plan should establish a record-keeping and reporting procedure for all events during and after the radiological emergency. It should also identify the person responsible for preparing the report (NRRC, 2023). Where appropriate, the following records should be kept:

- Emergency worker dose records
- Radiation monitoring and contamination level recording
- Radioactive waste produced
- Education
- Emergency exercise/practice and actual events
- Equipment calibration record

11. Recovery and Rehabilitation

Each emergency plan should address the characteristics of the facility, employees, or site following the emergency, as well as any identified recovery issues. The plan should define both short-term and long-term recovery actions to be taken on-site and off-site to restore services for employees, the public, and facilities affected by the radiological emergency back to serving the public should be defined in the plan (IAEA, 2020c). Rescue operations ensure that contaminated areas and materials are cleaned and reduced below natural levels through repeated measurement and pollution tests and decontamination processes. If the contamination level cannot be reduced to the targeted level or there is long half-life radionuclide contamination, the materials are treated as radioactive waste. Additionally, the rescue operation may include decontamination of the affected area and the surrounding area (NRRC, 2023).

Radioactive waste produced in this way must be managed in accordance with the requirements of the "Radioactive Waste Management Regulation" published by the national authority (NCRP, 2010).

12. Revision of the Radiation Emergency Plan

Lessons learned from emergency drills should be evaluated in the light of national and international experience, revisions in reference documents and daily needs, and the emergency plan should be revised if necessary.

13. Quality Assurance

Ensuring the reliability of the materials, measurement systems, equipment and communication necessary for executing the activities outlined in the radiation emergency plan is crucial. The most important step to achieve this is to establish a quality assurance system.

In this context, the quality assurance system to be established will ensure that the equipment is ready for use and functional during a possible emergency response by making arrangements for its inventories, materials, testing and calibration methods

14. Training and Exercises for Emergency Preparedness and Response

The management organises training and ensures that the personnel involved in the emergency response participate in drills so that they can effectively perform their duties in a radiological emergency.

The organisation should arrange training to ensure that the personnel selected for the response team acquire the knowledge, skills and abilities to ensure that they have the necessary qualifications. The program will include arrangements for continuous refresher training and to ensure that personnel authorised in emergency response undergo specified training (NCRP, 2010).

Although medical physicists working in hospitals generally possess sufficient knowledge and experience in radiation dosimetry, such as dose planning and dose measurements, they often lack the specific knowledge and skills required in nuclear or radiological emergencies. This is actually due to the fact that training that includes specific knowledge and skills for emergency situations is not provided within the curriculum during physics education.

This section proposes a training program that encompasses the information medical physicists need in nuclear or radiological emergencies. This training will serve a complementary role in preparing medical physicists effectively to respond to nuclear or radiological emergencies. This will also enable medical physicists to actively contribute to the training of other healthcare professionals in responding to nuclear or radiological emergencies. General educational aspects and content of specific topics of the training program for medical physicists (IAEA, 2020d; NYC, 2017; IAEA, 2018);

- To introduce the role of the medical physicist in effectively responding correctly to nuclear or radiological emergencies.
- To provide an overview of nuclear or radiological emergencies by presenting emergency response in general terms.
    - Overview of nuclear and radiological emergencies
    - Information about emergency types and emergency planning categories
    - General aspects of preparation for and response to nuclear and radiological emergencies;
    - Applicability of IAEA security standards to specific situations.
- To introduce the main types of measurement and survey devices used in nuclear and radiological emergencies and to ensure the acquisition of the necessary knowledge and skills to measure radiation levels with appropriate instrumentation.
    - Carry out radiation measurements and surveys using a range of equipment;
    - Carry out functional and calibration checks on measuring equipment;
    - Discuss the challenges of measuring alpha, beta, low energy photons and neutrons;
    - Select appropriate monitoring tools for nuclear or radiological emergencies;
    - Understanding the physical limitations and uncertainties of dose measurements;
    - Maintaining an inventory of existing equipment in the hospital;
    - Responding to radiological emergencies;
    - General information about radiation measurement devices, measurement capabilities, maintenance, and understanding of their use in nuclear or radiological emergencies.
- To understand the methods and tools that can be used for dose assessment and reconstruction in the context of medical management in nuclear or radiological emergencies.
    - Information on the ICRU-60 and ICRP-103 reports;
    - Basic information on internal dose calculation models;

    - Understanding of radiation measurement quantities and units required for dose assessment and accident scenario building.
    - Monitoring and decontamination of people at the scene and in reception centres, and waste management;
    - Information on reception centres;
    - Identify the roles and responsibilities of a medical physicist in a reception centre.
- Monitoring and Decontamination of People and Waste Management in Hospital;
    - Focusing on unique issues related to monitoring and decontamination in hospitals.
- Biological effects of radiation: Effects on cells and tissues, as well as stochastic effects;
    - Understanding the processes of carcinogenesis and the genetic effects of radiation and the factors affecting them;
    - Understand the effects of ionising radiation on cells and tissues and their medical consequences;
    - Classification of biological effects of radiation according to relevant dose assessment, dose-response and risk models;
- Public protection strategies; to provide an understanding of the basic concepts, strategies and measures for protecting the public in the event of a nuclear or radiological emergency.
- Worker protection strategies; to provide an understanding of the basic concepts, strategies and precautions for the protection of workers in nuclear or radiological emergencies.
- Medical management; general information on medical intervention in nuclear and radiological emergencies.

15. Decontamination Procedure

This section provides recommendations for the personnel affected by any radioactivity contamination occurring in the facility or activity area and for cleaning the contaminated area (NCRP, 2010; University of Georgia, 2021).

15.1. Radioactive Material Spill

A spill of radioactive material in a facility or activity area is deemed a major spill if the resulting contamination contains substances that could generate significant levels of airborne radioactivity (mist, dust, smoke). Any malfunction of radiation-producing X-ray equipment, accelerators, and closed-source teletherapy devices that have the potential to generate high radiation should be treated on the same level as a spill (NCRP, 2010; University of Georgia, 2021).

15.1.1. Major Spills

Precautions to be taken by personnel in case of a major spill (University of Georgia, 2021):

- Stop working and immediately secure safety hazards.
- Alert other employees in the area; personnel should evacuate and take measures to prevent the spread of pollution.
- Potentially contaminated persons should be radiologically monitored before leaving the area.
- Cordon the area to prevent spillage from spreading.
- If there are any volatile substances or the potential for radioactivity in the air, make sure fume cupboards are operating. Close existing doors to control ventilation. If you are outdoors, stand against the wind.
- Lock doors, post warning signs, and radiation protection personnel can perform radiological monitoring of the affected area from a safe distance.
- Notify radiation safety personnel. Stay in a safe place until help arrives.
- The relevant personnel should not leave the scene until it is cleared by emergency response personnel. (UNLV, 2018; University of Georgia, 2021).

15.1.2. Minor Spills

A minor spill is a spill that contains small amounts of radioactive material. Properly controlled small spills should not cause contamination of personnel. Precautions to be taken by personnel in case of a minor spill (University of Georgia, 2021):

- Stop working and immediately secure safety hazards.
- Warn people in the area to stay away from pollution.
- Inform radiation safety personnel if necessary.
- Secure the area to prevent spillage from spreading. Cover liquid spills with absorbent materials.
- Monitor people exposed to contamination as a result of the spill.
- Trained personnel wearing protective clothing (gloves, laboratory coat, etc.) should carefully remove contamination. Collect used absorbent materials in the radioactive waste container for disposal.
- If contamination is detected in the area, the area should be decontaminated according to the application in Section 16.2 or Radiation safety personnel should be contacted.
- Report the accident and forward a copy to the Radiation safety office.

15.1.3. Precautions for Pollution on Skin and Clothes

- If the spill is on clothing, the clothing should be removed immediately and the skin areas in contact with the spill should be monitored as described in section 16.3. When removing clothing, care should be taken not to spread contamination to other parts of the body.
- If the radioactive material is volatile, place the contaminated clothing in a work fume hood or plastic bag and tie tightly.
- If radioactive material is spilled directly on the skin, wash the affected area immediately with running water. Warm water is best. Cold water may cause the skin's pores to close, and the pores may retain contamination. Hot water can cause the pores to open, allowing the contaminant to penetrate the layers of skin.
- Clean the affected area with a disposable towel and continue to monitor for contamination according to the procedure in Section 16.3.
- Report any contaminated skin or personal clothing as suspected or confirmed radioactive contamination to radiation safety personnel immediately.

15.2. Decontamination Guide for Areas and Equipment

Perform decontamination of the area or equipment (UNLV, 2018; University of Georgia, 2021);

- Wear protective clothing, control access to the area and ensure that personal clothing does not come into contact with contaminated surfaces during decontamination.
- Determine the approximate boundaries of the contaminated area with a radiation meter or dirt sample test.
- Mark the boundaries of the contaminated area with tape or a similar method.
- Carefully clean the dirty area using cleaning materials and disposable towels. Do not use volatile solvents and excessive amounts of water.
- If sprayed cleaners are used, do not spray directly onto contaminated surfaces close to the surface. Violent spraying techniques may spread contamination.
- When wiping with disposable towels, it is usually beneficial to wipe the dirtiest section. Then it should be wiped from the less dirty area to the dirtier area. Make a single pass when wiping and use a new surface of the towel each time you wipe. Wiping in a circular motion, usually towards the centre, is effective in preventing the spread of contamination.
- Properly dispose of all waste. The wet contaminated towel should be placed in a dry radioactive waste container containing sufficient absorbent material to prevent the formation of visible liquid.
- Conduct contamination surveys and continue decontamination operations if necessary.
- Perform personnel contamination monitoring after each cleaning process.
- If at least three decontamination processes are unsuccessful, different decontamination agents or methods should be used. If necessary, seek assistance from a radiation protection specialist.
- As a result of radiological monitoring, the adequacy of decontamination is confirmed under conditions where contamination is within natural limits, and waste materials are disposed of in

accordance with the rules and documented by examinations (University of Georgia, 2021).

15.3. Contamination Monitoring Procedure in Personnel

- Turn the measuring range of the radiation meter to the lowest setting and allow the instrument to stabilise with respect to natural radiation. Monitoring of worker radiation pollution should be carried out in the area with the lowest natural radiation level.
- First of all, make sure your hands are clean to avoid spreading contamination.
- Slowly survey the surface under suspicion of contamination with the device's detector from a distance of approximately 2 cm.
- Check all areas that could be contaminated: the torso, arms, elbows, hands, head and feet, and clothing.
- If there is an increase in the counting rate of the measuring system audibly or, hold the detector stationary over this location for 5 to 10 seconds and determine if the reading is higher than the natural radiation level.
- If contamination is high, ask for assistance in notifying Radiation Safety personnel to stay where you are and prevent spread.
- If no contamination is detected, consider the need to survey other areas and monitor personnel.

15.3.1. Response to Personnel Contamination Incidents

- Always notify Radiation safety personnel if skin contamination is confirmed.
- Before starting decontamination, try to determine the location and approximate size of the contaminated area.
- Simple skin contamination can usually be removed by washing the affected area with mild soap and warm water. Make sure that the contamination does not spread to other parts of the body during the decontamination process.
- Try to clean the area by touching it with a disposable towel. Survey the affected area again after the procedure. Repeat decontamination with soap and water if necessary.

- If three consecutive decontamination procedures using mild soap and water are not successful, repeat decontamination should be performed with industrial-grade hand sanitizer.
- To avoid irritating the skin, do not use harsh chemicals or try to clean injured areas or eyes. Decontamination must be done with the support of medical experts and Radiation safety personnel.
- Contamination on the face, cracks in the skin and contamination with the potential to be absorbed by the skin, or internal contamination may require bioanalysis (University of Georgia, 2021).

Conclusion

This study outlines the organisational structure of the government and facilities for emergency response and prevention and the procedures to be followed in the event of a possible nuclear or radiological accident. It encompassess efforts to protect employees and the public from radiation, to reduce the harmful effects of radiation, and to restart facilities in a short time. The study defines the roles and responsibilities of the groups formed within the Emergency Response and Prevention organisational structure.

The Emergency Response organisation ensures that training is organised within the framework of each group's area of responsibility, guaranteeing that the selected personnel have the necessary qualifications and to acquire the knowledge, skills, and abilities to perform the intervention functions. Furthermore, the training program presented in the study includes arrangements for continuous refresher training and arrangements to ensure that personnel assigned to responsible positions in emergency response undergo the specified training.

Additionally, this work will ensure coordination with international organisations, foster bilateral relations for radiological and nuclear emergency response and prevention, and enhance international emergency response capability.

References

IAEA, International Atomic Energy Authority. (1998). Planning the Medical Response to Radiological Accidents, Safety Reports Series No. 4, IAEA, Vienna.

IAEA, International Atomic Energy Authority. (2006a). Manual for First Responders to a Radiological Emergency, Emergency Preparedness and Response EPR-First Responders, IAEA, Vienna.

IAEA, International Atomic Energy Authority. (2006b). Environmental Consequences of the Chernobyl Accident and Their Remediation: Twenty Years of Experience, Radiological Assessment Reports Series, IAEA, Vienna.

IAEA, International Atomic Energy Authority. (2007). Arrangements for Preparedness for a Nuclear or Radiological Emergency, IAEA Safety Standards Series No. GS-G-2.1, IAEA, Vienna.

IAEA, International Atomic Energy Authority. (2011). Criteria for Use in Preparedness and Response for a Nuclear or Radiological Emergency, IAEA Safety Standards Series No. GSG-2, IAEA, Vienna.

IAEA, International Atomic Energy Authority. (2012). Lessons Learned from the Response to Radiation Emergencies, 1945-2010. EPR-Lessons Learned 2012, IAEA, Vienna.

IAEA, International Atomic Energy Authority. (2014). Radiation Protection and Safety of Radiation Sources: International Basic Safety Standards, IAEA Safety Standards Series No. GSR Part 3, IAEA, Vienna.

IAEA, International Atomic Energy Authority. (2015). Preparedness and Response for a Nuclear or Radiological Emergency, IAEA Safety Standards Series No. GSR Part 7, IAEA, Vienna.

IAEA, International Atomic Energy Authority. (2018). Radiation Protection and Safety in Medical Uses of Ionizing Radiation, IAEA Safety Standards Series No. SSG-46, IAEA Vienna.

IAEA, International Atomic Energy Agency. (2020a). Medical Management of Radiation Injuries, Safety Reports Series No. 101, IAEA, Vienna.

IAEA, International Atomic Energy Authority. (2020b). Considerations in the development of a protection strategy for a nuclear or radiological emergency, Emergency Preparedness and Response (EPR), Vienna.

IAEA, International Atomic Energy Authority. (2020c). Guidance for Medical Physicists Responding to a Nuclear or Radiological Emergency, Emergency Preparedness and Response (EPR) IAEA, Vienna.

IAEA, International Atomic Energy Authority. (2020d). Public Communication in Preparedness and Response for a Nuclear or Radiological Emergency, IAEA Safety Standards Series No. GSG-14, IAEA, Vienna.

NCRP, National Council on Radiation Protection and Measurements REPORT No. 165. (2010). Responding to a Radiological or Nuclear Terrorism Incident: A Guide for Decision Makers, Recommendations of the National Council on Radiation Protection and Measurements, NCRP Report No. 165.

NRRC, Stakeholders Guidelines. (2023). Development of Emergency Plan for Radiological Facilities Stakeholder Guideline NRRC-SG-018.

NYC, Department of Health and Mental Hygiene (DOHMH). (2009). NYC Health. Hospital Guidance for Responding to a Contaminating Radiation Incident, Healthcare Emergency Preparedness Program, New York City,

https://www.nyc.gov/assets/doh/
downloads/pdf/bhpp/hepp-radmanualpt 1-042309.pdf.
Pedraza, J. M. (2013). World major nuclear accidents and their negative impact in the environment, human health and public opinion. International Journal of Energy, Environment and Economics, 21(2), 1-23.
PNRA, Pakistan Nuclear Regulatory Authority. (2016). Preparation of Radiation Emergency Plan for Radiation Facilities and Activities, Regulatory Guide, Pakistan Nuclear Regulatory Authority, PNRA-RG-914.02.
University of Georgiya, Radiation Safety Manual. (2021). Radiation Safety Program 706-542-0107. https://research.uga.edu/docs/units/safety/ radiation-forms/Chapter_8.pdf.
UNLV, University of Nevada, Las Vegas. (2018). Radiation Emergency Response Plan, Risk Management & Safety University of Nevada, Las Vegas, Nevada.

Index

D

E

F

L

M

N

O

P

Q

R